ELECTRICIAN'S
Technical Reference

Motors

ELECTRICIAN'S
Technical Reference

Motors

David Carpenter

DELMAR
CENGAGE Learning

Australia • Brazil • Japan • Korea • Mexico • Singapore • Spain • United Kingdom • United States

Electrician's Technical Reference: Motors
David Carpenter

Publisher: Alar Elken

Acquisitions Editor: Mark Huth

Developmental Editor: Jeanne Mesick

Production Manager: Larry Main

Art Director: Nicole Reamer

Editorial Assistant: Dawn Daugherty

Cover Design: Nicole Reamer

For product information and technology assistance, contact us at
Cengage Learning Customer & Sales Support, 1-800-354-9706

For permission to use material from this text or product,
submit all requests online at **cengage.com/permissions**
Further permissions questions can be emailed to
permissionrequest@cengage.com

Library of Congress Control Number: 98-17357

ISBN-13: 978-0-8273-8513-9

ISBN-10: 0-8273-8513-7

Delmar
Executive Woods
5 Maxwell Drive
Clifton Park, NY 12065
USA

Cengage Learning is a leading provider of customized learning solutions with office locations around the globe, including Singapore, the United Kingdom, Australia, Mexico, Brazil, and Japan. Locate your local office at: **international.cengage.com/region**

Cengage Learning products are represented in Canada by Nelson Education, Ltd.

For your lifelong learning solutions, visit **delmar.cengage.com**

Visit our corporate website at **www.cengage.com**

Notice to the Reader

Publisher does not warrant or guarantee any of the products described herein or perform any independent analysis in connection with any of the product information contained herein. Publisher does not assume, and expressly disclaims, any obligation to obtain and include information other than that provided to it by the manufacturer. The reader is expressly warned to consider and adopt all safety precautions that might be indicated by the activities described herein and to avoid all potential hazards. By following the instructions contained herein, the reader willingly assumes all risks in connection with such instructions. The publisher makes no representations or warranties of any kind, including but not limited to, the warranties of fitness for particular purpose or merchantability, nor are any such representations implied with respect to the material set forth herein, and the publisher takes no responsibility with respect to such material. The publisher shall not be liable for any special, consequential, or exemplary damages resulting, in whole or part, from the readers' use of, or reliance upon, this material.

Printed in the United States of America
7 8 9 10 11   15 14 13 12 11

FD306

Contents

Preface

Everyone responsible for industrial and commercial facilities knows the reliability of any facility depends on its motors. It is almost impossible to imagine any facility, process, or entity that does not involve electric motors somewhere or somehow.

It is obvious that proper design, understanding and maintenance of motors is essential to the dependability of your operation.

When an essential motor goes down in a high-production plant or a process-sensitive operation, lost production and lost revenues may run into hundreds of thousands of dollars. Failure of a simple conveyor motor can hold up an entire plant and idle thousands of workers. Downtime is costly in a variety of processes—pharmaceutical, steel, heavy machinery, semiconductor manufacturing, and laboratory research work. Machines that incorporate several motors and operate at less than 100 percent show a percentage of loss revenue that can amount to a large financial loss when left unrepaired for long periods of time.

Another cost occurs when those performing the maintenance, design, and installations of motors do not understand the proper operation and application of motors. This problem may go unnoticed for some time, but will result in a large net loss.

This book is written with the electrician, instrument mechanic, and engineer in mind. It should prove to be a valuable resource for those involved with motor application, troubleshooting, and training. The text will explain the charts, tables, and illustrations that are to be used as a quick references for field applications.

The text is designed to build on one concept at a time. Some may skip the first two chapters and skip to the troubleshooting guides and techniques. Others may use Chapters 2 and 10 to help design or redesign their motor system. Chapter 1 gives definitions to the nameplate information provided by the manufacturers. Understanding this information is essential to properly installing and troubleshooting motors. Chapter 2 describes how to properly install and design the wiring method associated with the motor. This is also critical because the misapplication of wiring methods is largely responsible for the breakdown of motors. Chapter 3 illustrates motor theory and how to apply it to field situations. Chapters 4 through 8 specify types of single- and three-phase motors. These chapters also provide troubleshooting guides and techniques for specific type motors. Chapter 9 gives test standards for testing the motor and performing maintenance. Chapter 10 provides information on how to make your motor system operate at greatest efficiency. The charts, illustrations, and tables are quick reference guides that should prove useful for those working with motors.

Acknowledgments

The author and Delmar Publishers would like to thank the following reviewers for the comments and suggestions they offered during the development of this project. Our gratitude is extended to:

Jeff Auter
Kirby Risk
Lafayette, Indiana

Kevin Earley
Bay Harbour Electric
Erie, Pennsylvania

Wayne Sorge
Independent Electric Contractors
Denver, Colorado

We would also like to acknowledge and thank Siemens Energy and Automation, Inc. for allowing us generous use of many illustrations.

Introduction to Motors: Design and Maintenance

Motors are used worldwide in many residential, commercial, industrial, and utility applications. Motors transform electrical energy into mechanical energy. A motor may be part of a pump or fan, or may be connected to some other form of mechanical equipment such as a winder, conveyor, or mixer. Motors are found on a variety of applications, from single motor applications to several motors.

Understanding Nameplate Information for Design, Installation, and Maintenance

Much of the information needed to make the proper selection, installation, and maintenance of a motor is found in the nameplate information. According to the National Electrical Code (NEC), Article 430, the nameplate of a motor must contain the following information (see Figure 1–1).

Manufacturer's Name

The name tells the customer who manufactures the motor. It is not essential to replace the motor with a motor from the same manufacturer. Other information on the nameplate will help you to order the proper replacement.

Type

Type is a letter designation by the manufacturer. This can tell you the style of the motor (for example, split-phase, capacitor start-induction run). This information is critical if a suitable replacement is to be found.

RPM

Base speed is the nameplate speed, given in revolutions per minute (RPM), where the motor develops rated horsepower at rated voltage and frequency. It is an indication of how fast the output shaft will turn the connected equipment when fully loaded with proper voltage and frequency applied. The base speed of a 1765-RPM, 460-volt, three-phase motor at 60 Hz, with a synchronous speed of a four-pole motor, is 1800 RPM. When fully

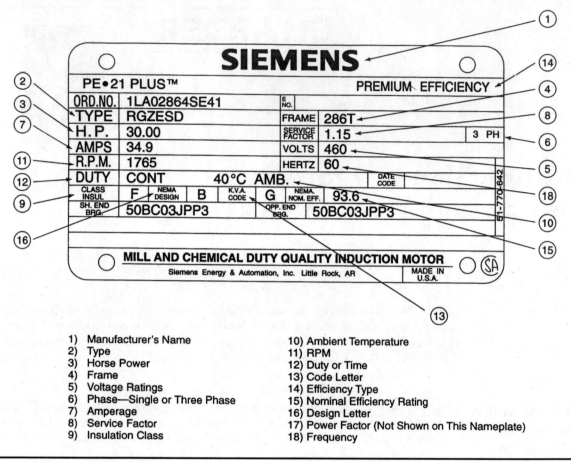

1) Manufacturer's Name
2) Type
3) Horse Power
4) Frame
5) Voltage Ratings
6) Phase—Single or Three Phase
7) Amperage
8) Service Factor
9) Insulation Class
10) Ambient Temperature
11) RPM
12) Duty or Time
13) Code Letter
14) Efficiency Type
15) Nominal Efficiency Rating
16) Design Letter
17) Power Factor (Not Shown on This Nameplate)
18) Frequency

Figure 1–1   Nameplate information; Courtesy Siemens Energy and Automation, Inc.

loaded, there will be 1.9 percent slip. If the connected equipment is operating at less than full load, the output speed (RPM) will be slightly greater than the nameplate speed.

$$\% \text{ Slip} = 1800 - 1765 / 100$$

$$\% \text{ Slip} = 1.9\%$$

Horsepower

This is the rated mechanical horsepower (HP) output of the motor. It is measured at the rated voltage and current and at the proper applied frequency.

Frame

If the motor was manufactured under National Electrical Manufacturers Association (NEMA) guidelines, the frame number gives all the critical measurements of the motor (see Figure 1–2).

NEMA has *standardized frame size* motor dimensions. Standardized dimensions include bolt hole size, mounting base dimensions, shaft height, shaft diameter, and shaft length.

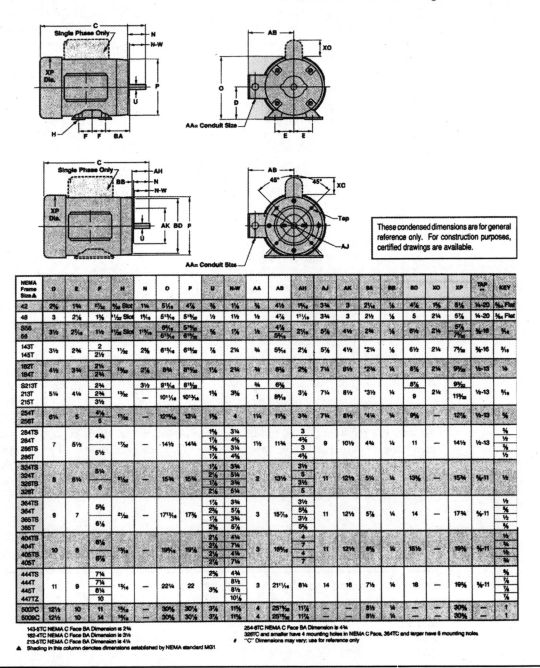

Figure 1–2 Frame size and dimension data; Courtesy Keljik, *Electric Motors and Motor Controls,* Delmar Publishers.

Existing motors can be replaced without reworking the mounting arrangement. New installations are easier to design because the dimensions are known. Letters are used to indicate where a dimension is taken. For example, the letter C indicates the overall length of the motor. The letter E represents the distance from the center of the shaft to the center of the mounting holes in the feet. You can determine the actual dimensions by referring to a table in the motor data sheet and referencing the letter (see Fig. 1–3a).

NEMA divides standard frame sizes into two categories—*fractional* and *integral.* Fractional frame sizes are designated 48 and 56. They primarily include horsepower ratings of less than one horsepower. Integral- or medium-horsepower motors are designated by frame sizes ranging from 143T to 445T. A T in the motor frame-size designation of integral horsepower motors indicates that the motor is built to current NEMA frame standards. Motors built prior to 1966 have a U in the motor frame-size designation, indicating they are built to previous NEMA standards.

143T = current NEMA standards

326U = previous NEMA standards

The frame-size designation is a code to help identify key frame dimensions. The first two digits, for example, are used to determine the shaft height. The shaft height is the distance from the center of the shaft to the mounting surface. To calculate the shaft height, divide the first two digits of frame size by 4. In Figure 1–3b, a 143T frame-size motor has a shaft height of 3.50 inches (14 ÷ 4).

The third digit in the integral T frame-size number is the NEMA code for the distance between the center lines of the mounting-bolt holes in the feet of the motor (Figure 1–3c). The dimension is determined by matching the third digit in the frame number with a table in NEMA publication MG-1. From this we determine that the distance between the center lines of the mounting-bolt holes in the feet of a 143T frame is 4.00 inches (see Table 1–1).

International European Commission also has standardized dimensions, which differ from NEMA. Many motors are manufactured using International European Commission (IEC) dimensions. IEC dimensions are shown in Figure 1–3d.

The typical *floor-mounting positions* are illustrated in Figure 1–3e , and are referred to as F-1 and F-2 mountings. The conduit box can be located on either side of the frame to match the mounting arrangement and position. The standard location of the conduit box is on the left-hand side of the motor when viewed from the shaft end. This is the F-1 mounting. The conduit opening can be placed on any of the four sides of the box by rotating the box in 90 degree steps.

TABLE 1.1   How to Determine Mounting-Hole Distance

Frame Third/Fourth Digit In Frame Number

Number Series	D	1	2	*3*	4	5
140				*4.00*		
160	4.00	3.50	4.00	4.50	5.00	5.00
180	4.50	4.00	4.50	5.00	5.50	5.50
200	5.00	4.50	5.00	5.50	6.50	6.50
210	5.25	4.50	5.00	5.50	6.25	6.25
220	5.50	5.00	5.50	6.25	6.75	6.75
250	6.25	5.50	6.25	7.00	8.25	8.25
280	7.00	6.25	7.00	8.00	9.50	9.50
320	8.00	7.00	8.00	9.00	10.50	10.50

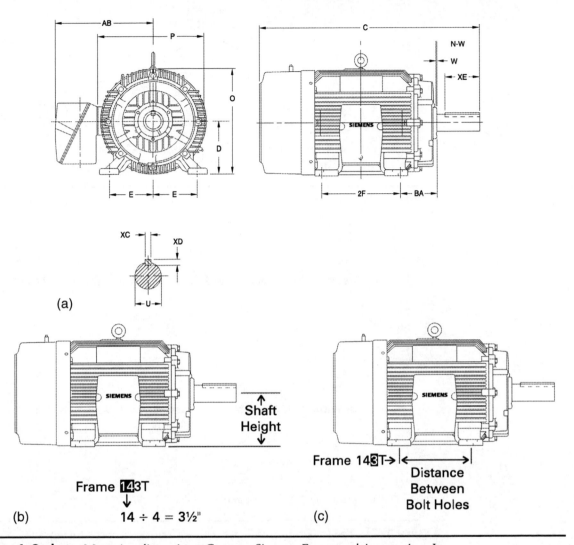

Figure 1–3a,b,c   Mounting dimensions; Courtesy Siemens Energy and Automation, Inc.

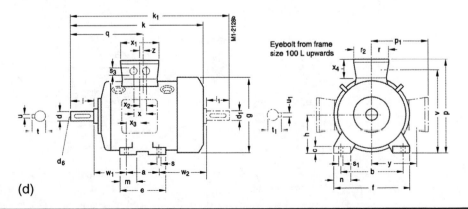

Figure 1–3d   IEC dimensions; Courtesy Siemens Energy and Automation, Inc.

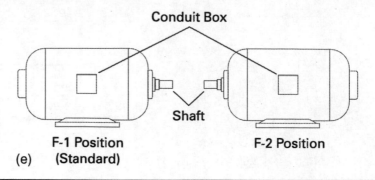

F-1 Position
(e) (Standard)

F-2 Position

Figure 1–3e Mounting dimension; Courtesy Siemens Energy and Automation, Inc.

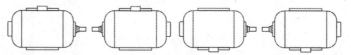

Assembly W-1 Assembly W-2 Assembly W-3 Assembly W-4

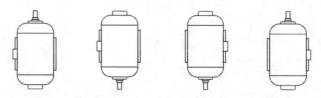

Assembly W-5 Assembly W-6 Assembly W-7 Assembly W-8

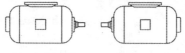

(f) Assembly C-1 Assembly C-2

Figure 1–3f Foot mounting, wall and ceiling; Courtesy Siemens Energy and Automation, Inc.

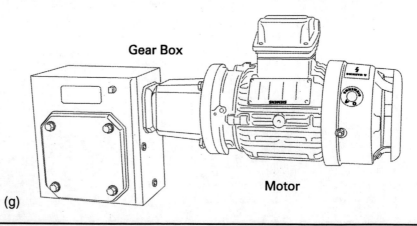

Gear Box

Motor

(g)

Figure 1–3g Mounting faces; Courtesy Siemens Energy and Automation, Inc.

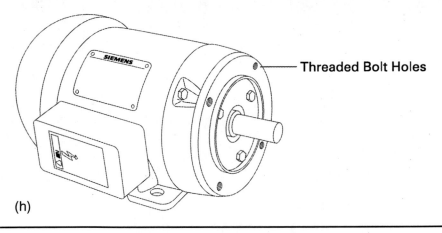

(h)

Figure 1–3h   C-face mounting; Courtesy Siemens Energy and Automation, Inc.

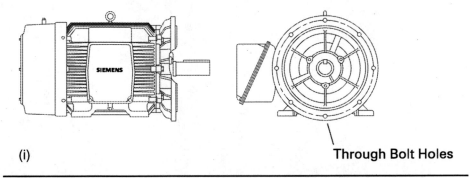

(i)

Figure 1–3i   D-flange mounting; Courtesy Siemens Energy and Automation, Inc.

With modification, the *foot-mounted motor* can be mounted on the wall or ceiling. Typical wall and ceiling mounts are shown in Figure 1–3f . Wall-mounted positions have the prefix W and ceiling-mounted positions have the prefix C. It is sometimes necessary to connect the motor directly to the equipment it drives. In Figure 1–3g, a motor is connected directly to a gear box. The face, or the end, of a *C-face motor* has threaded bolt holes. Bolts to mount the motor pass through mating holes in the equipment and into the faces of the motor. (see Figure 1–3h). In a *D-flange motor,* the bolts go through the holes in the flange and into threaded mating holes of the equipment (see Figure 1–3i).

Vertical Pump Motors   *Vertical hollow-shaft pump motors* are designed for vertical pump applications. The motors are squirrel-cage induction type with NEMA design B torque and current characteristics. Motors are rated from 25 to 250 HP and 1800 RPM. Vertical pump motors are designed for 460-volt, three-phase, 60-Hz systems. Thermostats and space heaters are optional (see Figure 1–4). These motors have a NEMA standard P-flange mounting shaft with a hollow shaft that accommodates the driven shaft to extend through the rotor. The coupling for connecting the motor shaft to the driven shaft is located in the top of the motor. Solid-shaft motors, are more conventional than hollow-shaft motors. The motor shaft for the solid-shaft motors is coupled to the driven shaft below the P-flange face.

Figure 1–4   Vertical pump motor; Courtesy Siemens Energy and Automation, Inc.

Vertical solid-shaft motors designed for in-line pump applications are available from 3 to 100 HP at 3600 RPM, and 3 to 250 HP at 1200 and 1800 RPM.

Above NEMA Motors   Motors that are larger than the NEMA frame sizes are referred to as *above NEMA motors*. These motors typically range in size from 200 to 10,000 HP. There are no standardized frame sizes or dimensions, because above NEMA motors are typically constructed to meet the specific requirements of an application (see Figure 1–5).

Volts

This value is the voltage to be delivered to the motor terminals. If it is a dual-voltage motor, either voltage may be applied, but internal connections must be changed. Several factors can affect the operation and performance of an AC motor. These need to be considered when applying a motor.

AC motors are designed to operate on standardized voltages and frequencies. Table 1–2 reflects NEMA standards.

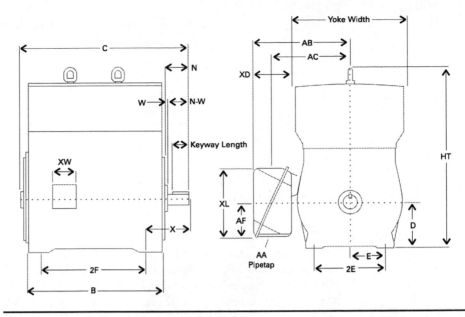

Figure 1–5 NEMA Motors—Motors with larger than NEMA size frames; Courtesy Siemens Energy and Automation, Inc.

TABLE 1.2 **Standard Voltage and Frequencies**

60 Hz	50 Hz
1 15 VAC	380 VAC
200 VAC	400 VAC
230 VAC	415 VAC
460 VAC	220/380 VAC
575 VAC	

TABLE 1.3 **Affect of Frequency Variation on Speed and Torque**

Frequency Variation	% Change Full-Load Speed	% Change Starting Torque
+5%	+5%	–10%
–5%	–5%	+11%

A small variation in supply voltage can dramatically affect motor performance. In Table 1–3, for example, when voltage is 10 percent below the rated voltage of the motor see Table 1–2, the motor has 20 percent less starting torque. This reduced voltage may prevent the motor from getting its load started or keeping it running at rated speed. A 10 percent increase in supply voltage, on the other hand, increases the starting torque by 20 percent. This increased torque might cause damage during startup. A conveyor, for example, might lurch forward at startup. A voltage variation will cause similar changes in the motor's starting amps, full-load amps, and temperature rise.

Frequency

A variation in the frequency at which the motor operates causes changes primarily in speed and torque characteristics as shown in Table 1–3. A 5 percent increase in frequency, for example, causes a 5 percent increase in full-load speed and a 10 percent decrease in torque.

TABLE 1–4 **Insulation Class Ratings**

CLASS A: 105° C
CLASS B: 130° C
CLASS F: 155° C
CLASS H: 180° C
CLASS H+: 200° C

Phase

Either single- or three-phase will be designated. *Note:* A three-wire service to a home does not constitute three-phase power.

Amperage

This is normal operating current when running at rated horsepower and rated voltage. If two currents are listed, the higher current is associated with the lower voltage connection and the lower current is associated with the higher voltage.

Note: Running currents for single-phase motors are given in NEC's Table 430–148. You will notice that these currents may not correspond to the actual motor nameplate ratings. The code book values are used to size several components of the motor system and are used as a guide for generic motor calculations.

Service Factor (SF)

Service factor of the motor is a multiplier applied to the motor horsepower. It indicates how much the motor can run overloaded on a continuous basis without damaging the motor winding insulation.

For example, a motor designed to operate at its nameplate horsepower rating has a service factor of 1.0. This means the motor can operate at 100 percent of its rated horsepower. Some applications may require a motor to exceed the rated horsepower. In these cases, a motor with a service factor of 1.15 can be specified. A 1.15 service factor motor can be operated 15 percent higher than the motor's nameplate horsepower. Thus, a 1.15 SF multiplied to a 1 HP motor can safely run at 1.15 HP without damage at listed ambient temperature or lower.

It should be noted that any motor operating continuously at a service factor greater than 1.0 will have a reduced life expectancy compared to operating it at its rated horsepower. In addition, performance characteristics such as full-load RPM and full-load current will be affected.

Insulation Class

The class of insulation used on the motor windings indicates the maximum operating temperature of the coil windings in the motor (see Table 1–4).

Ambient Temperature

If listed, the ambient temperature is the maximum temperature in degrees centigrade of the surrounding air, which will allow rated horsepower without damage. (*Note:* The revolutions per minute is the rated speed of the motor operating at full load or rated HP. Typically, lighter loads will run at higher speeds.)

TABLE 1–5   **Typical Horsepower Derating Factors**

Altitude	Derating Factor
3300–5000	0.97
5001–6600	0.94
6601–8300	0.90
8301–9900	0.86
9901–11,500	0.82

Duty (or Time Rating)

Duty refers to the length of time the motor can run at full load without overheating. Continuous duty means that the motor could run 24 hours a day if all other factors are within specifications.

Code Letter

Locked rotor KVA per horsepower, if 1/2 horsepower, or more, is listed in the NEC®. The code letter refers to the starting or locked rotor characteristics of the motor. It is listed as volt-amps per horsepower, with locked rotor. NEC Article 430–7b is a reference.

Efficiency

Efficiency is sometimes placed on the motor's nameplate. It is expressed as a percent of output watts (horsepower watts) compared to input watts at full-load conditions. For example: 85 percent efficient means that 85 percent of the input electrical watts are converted to mechanical output watts or horsepower.

Power Factor

The ratio of true power used in watts and the apparent power delivered is the power factor. It is expressed as a percentage.

Altitude

Motors are typically designed for altitudes below 3300 feet. Air cooling becomes a factor at higher altitudes, and derating of the motor's capabilities would be necessary. (Altitude is not normally indicated on the motor's nameplate.) Most motors must be derated for altitude changes. Table 1–5 gives typical horsepower derating factors, but the derating factor should be checked for each motor. A 50-HP motor operated at 6000 feet, for example, would be derated to 47 HP, providing the 40° C ambient rating is still required.

$$50 \text{ HP} \times 0.94 = 47 \text{ HP}$$

The ambient temperature may be reduced from 40° C to 30° C at 6600 feet on many motors. A motor with a higher insulation class may not require derating in these conditions.

Ambient Temperature (°C)	Maximum Altitude (Feet)
40	3300
30	6600
20	9900

Thermal Protection

If a motor has integral (internal to the motor) overload protection, it will be marked *thermally protected.* If 100w or less, the nameplate may says "(TP)."

Impedance Protection

If motors correspond to the impedance-protected definition NEC Article 430–32(c) (4) ratings, they may be marked "impedance protected" or "(ZP)" or "(imp.prot.)."

NEMA Design Letter

The design letter gives application that the motor should be applied. Design letters are classified A, B, C, D, and E. Figure 1–6 shows the speed torque curves for typical AC induction motors. Design letters are not the same as code letters. Design letters are intended to indicate operating characteristics, where code letters are used to size equipment based on the locked rotor KVA per horsepower.

Design A   Design A motors have relatively constant speed, or very low percent speed regulation. They have high starting torque and rapid acceleration, but higher starting current. The breakdown torque also is one of the highest at approximately 275 percent.

Design B   Design B motors are general-purpose design motors. They have lower starting torque than design A, but also have lower starting current. Breakdown torque is somewhat lower. Breakdown torque and starting current, are higher on Design A motors.

Design C   Design C motors have high starting torque and low starting current. Design C motors use a double squirrel cage to produce higher starting torque but lower starting current. The motor uses the high impedance of one squirrel cage winding located deep in the slots to keep current low during starting.

   The outer bars are small conductors with high resistance but low reactance. During starting, most of the current flow will be in the outer high-resistance bars. The phase relation between rotor and stator is nearly in phase because of the resistance of the bars. This

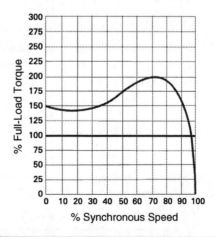

Figure 1–6   Speed and torque characteristics of design B motor; Courtesy Siemens Energy and Automation, Inc.

allows the magnetic fields to occur at the same time and to produce a high torque without a high current. As the rotor speed increases, the XL of the inner bars decreases as rotor frequency decreases. The inner windings are lower resistance and now also lower reactance, so the combined impedance is lower than the outer bars. Now rotor current flows in the inner bars. This allows the motor to react quickly to load changes, so it has good running torque and high breakdown torque. Of course, this motor costs more to purchase.

Design D   Design D motors use a high-resistance rotor to produce high starting torque, but they also have a high slip. These motors are typically suited to loads that drive high-inertia loads, such as flywheel loads. Starting current is high, so these motors should not be used for equipment requiring frequent starting. Design D motors are characterized by their high starting torques, combined with high slip. Starting torque for four-, six- and eight-pole design motors is 275 percent or more of full-load torque. Applications using stored energy flywheels are best matched with design D motors, reducing shock to both.

Design E   Design E motors are similar to the design A motor in that they have (1) a specific minimum efficiency; (2) limits on starting current; and (3) very similar starting torques. The technology and the application for design E and A is the same. Design E motors were developed to provide higher efficiency, but with this comes a higher locked rotor current, thus effecting overcurrent application. To date, no design E motors have been developed for commercial application—they are only in an experimental stage. Experiments have determined that the design E motor's locked rotor torque ratings decrease no more than 5 percent in motors 7 1/2 HP and above.

Different types of single phase motors have different applications. Not all motors will function satisfactorily in all situations. By understanding nameplate information, suitable replacement motors or original design motors can be more precisely selected. For instance, if a motor is not functioning as originally specified, the motor application may be wrong. Your knowledge of motor nameplates will allow you to choose the proper motor for the job.

Case Study: Using Nameplate Information

A three-phase motor for plant operations has been replaced three times in three years. The maintenance supervisor and engineer decide to look for a reason for the frequent need for replacement. After investigating the problem further, the supervisor notices that the nameplate rating shows that this motor is rated to be in an area of no greater than 150 degrees of ambient temperature. The ambient temperature at the motor location is a constant 195 degrees, and at time goes as high as 300 degrees. Upon further investigation, it was noted that the motor was installed several years prior to the installation of a new foundry. This motor was used as a pump. With installation of the new foundry the temperature had risen greater than what the original pump motor was rated. Failure to check the nameplate resulted in the loss of production and the cost to replace three motors.

Other parameters of the motor also give an indication of misuse or misapplication. Is the service factor (SF) critical for exact replacement of a defective motor? Does an SF of 1.15 fit the needs for replacement of a motor with an SF of 1.0? Normally, this would be acceptable if all other factors are considered, including the sizing of the overcurrent protection. Can a motor with higher insulation class replace one of lower class? Yes, this is acceptable; however, costs are a factor. Higher insulation class usually costs more for the same HP motor. Can duty rating be substituted? Can an intermittent duty motor replace a continuous duty motor? Check the application to see if the motor does, or could, run continuously.

Frames of the motor also are important to note. Open-frame motors need good air circulation to maintain operations without damage. If a motor is in a poorly ventilated area, replacement with the same type of motor will not prevent future failure. Often other changes must be made, such as using drip-proof motor housing, totally enclosed fan cooled, or explosion-proof motors. Make sure you know which motor frame style you need to satisfy the installation.

Motor Design*

Standard Motor Designs

Motors are designed with certain speed–torque characteristics to match speed–torque requirements of various loads. The five standard National Electrical Manufactures Association (NEMA) designs are NEMA A, NEMA B, NEMA C, NEMA D, and NEMA E. NEMA A is not used very often. NEMA B is most commonly used. NEMA C and NEMA D are used for specialized applications. A motor must be able to develop enough torque to start, accelerate, and operate a load at rated speed. Using the sample 1765-RPM, 30-HP motor discussed previously, torque (T) can be calculated by transposing the formula for horsepower.

$$HP = T \times RPM \div 5250$$
$$T = HP \times 5250 \div RPM$$
$$T = 30 \times 5250 \div 1765$$
$$T = 89.2 \text{ ft lb}$$

Speed–Torque Curve for NEMA B Motor

A graph like the one in Figure 1–6 shows the relationship between the speed and torque that the motor produces from the moment of start until the motor reaches full-load torque at rated speed. Figure 1–6 represents a NEMA B motor.

Starting Torque (Locked Rotor Torque)

Starting torque (point A on the graph) is also referred to as locked rotor torque. This torque is developed when the rotor is held at rest with rated voltage and frequency applied. This condition occurs each time a motor is started. When rated voltage and frequency are applied to the stator, there is a brief amount of time before the rotor turns. At this instant a NEMA B motor develops approximately 150 percent of its full-load torque. A 1765-RPM, 30-HP motor, for example, will develop approximately 133.8 lb-ft of torque (see Figure 1–7).

Accelerating and Breakdown Torque

The magnetic attraction of the rotating magnetic field will cause the rotor to accelerate. As the motor picks up speed, torque decreases slightly until it reaches point B on the graph. As speed continues to increase from point B to point C, torque increases until it reaches its maximum at approximately 200 percent. This torque is referred to as *accelerating* or *pull-up* torque. Point C is the maximum torque a motor can produce. At this point a 30-HP motor will develop approximately 178.4 lb-ft of torque. If the motor were overloaded beyond the motor's torque capability, it would stall or abruptly slow down at this point. This is referred to as *breakdown* or *pull-out* torque (see Figure 1–8).

*Written with permission from Siemens Energy and Automation, Inc.

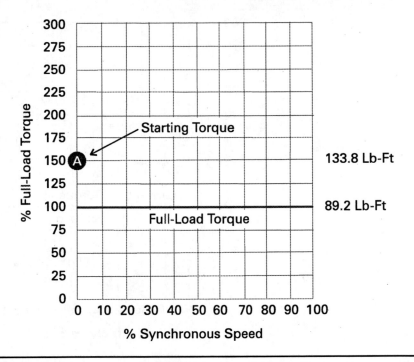

Figure 1–7 Starting torque and full-load torque; Courtesy Siemens Energy and Automation, Inc.

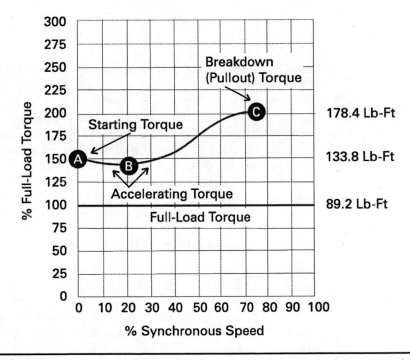

Figure 1–8 Accelerating and breakdown torque; Courtesy Siemens Energy and Automation, Inc.

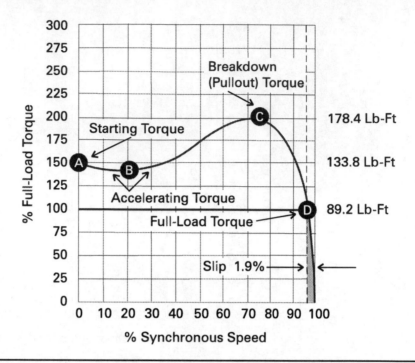

Figure 1–9   Full-load torque; Courtesy Siemens Energy and Automation, Inc.

Full-Load Torque

Torque decreases rapidly as speed increases beyond breakdown torque (point *C*), until it reaches full-load torque at a speed slightly less than 100 percent synchronous speed. Full-load torque is the torque developed when the motor is operating with rated voltage, frequency, and load. The speed at which full-load torque is produced is the slip speed or rated speed of the motor. Recall that slip is required to produce torque; this will be discussed later in the chapter. If the synchronous speed of the motor is 1800 RPM and the amount of slip is 1.9 percent, the full-load rated speed of the motor is 1765 RPM. The full-load torque of the 1765-RPM 30-HP motor is 89.2 ft lb. NEMA design B motors are general-purpose, single-speed motors suited for applications that require normal starting and running torque such as conveyors, fans, centrifugal pumps, and machine tools (see Figure 1–9).

Starting Current and Full-Load Current

Starting current is also referred to as *locked rotor current,* and is measured from the supply line at rated voltage and frequency with the rotor at rest. Full-load current is the current measured from the supply line at rated voltage, frequency, and load with the rotor up to speed. Starting current is typically 600 to 650 percent of full-load current on a NEMA B motor. Starting current decreases to rated full-load current as the rotor comes up to speed (see Figure 1–10).

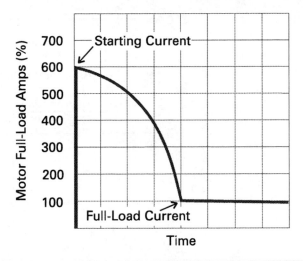

Figure 1–10   Starting current and full-load current; Courtesy Siemens Energy and Automation, Inc.

Design A Motor

NEMA A sets limits of starting (locked rotor) current for NEMA design B motors. When special load torque or load inertia requirements result in special electrical designs that will yield higher locked rotor current (LRA), NEMA design A may result. This designation also cautions the selection of motor control components to avoid tripping protective devices during longer acceleration times or higher than normal starting current.

Design B Motor

See Figure 1–6 for an example of a design B motor.

Design C Motor

Starting torque of a NEMA design C motor is approximately 225 percent. A NEMA C, 1765-RPM, 30-HP motor will develop approximately 202.5 ft lb of starting torque. Hard-to-start applications such as plunger pumps, heavily loaded conveyors, and compressors require this higher starting torque. Slip and full-load torque are about the same (see Figure 1–11).

Design D Motor

The starting torque of a NEMA design D motor is approximately 280 percent of the motor's full-load torque. A design D, with a full-load rated speed of 1765-RPM, 30-HP motor will develop approximately 252 ft lb of starting torque. Very-hard-to-start applications—such as punch presses, cranes, hoists, and oil-well pumps—require this high starting torque. Design D motors have no true breakdown torque. After initial starting torque is reached, torque decreases until full-load torque is reached. NEMA D motors typically are designed with 5 to 8 percent slip or 8 to 13 percent slip (see Figure 1–12).

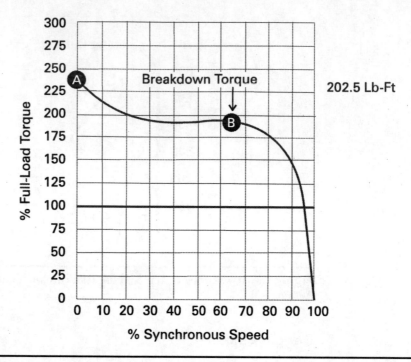

Figure 1–11 NEMA C full-load current and synchronous speed; Courtesy Siemens Energy and Automation, Inc.

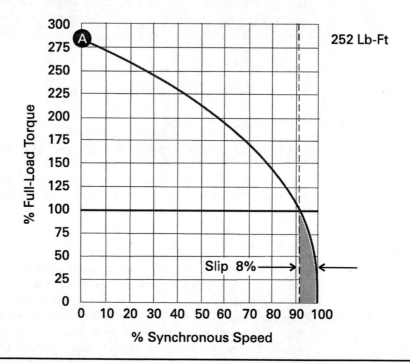

Figure 1–12 NEMA D motor full-load and current and synchronous speed; Courtesy Siemens Energy and Automation, Inc.

Design E Motor

The design E standard requires

1. The efficiency levels to be 1/2 to 2 points above the 12-6C federal legislation
2. Starting torque and current to be equivalent to the IEC Metric Motors of similar power ratings
3. The frame size and ratings to be equal to the NEMA design A and B motors.

Full-load torque runs over 300 percent at 85 to 90 percent of synchronous speed. Allowable starting current by the IEC is typically 15 percent to 20 percent higher than NEMA motors.

Multispeed and Adjustable Speed Drive (ASD)

These specialized motor designs are uniquely designed or selected to specific load requirements. NEMA design classifications are not applicable to these specialized motors.

Soft Starts

Various special configurations of motor controls are selected when starting/accelerating torques must be more accurately controlled, or when starting current must be limited. In the cases of part-winding start or wye-delta start, the motor windings must be designed with unique connections for the special controls. In cases such as reduced voltage auto-transformer or electronic soft starts, relatively standard motors may be approved for these special applications.

Motor Operation Categories

Motors consist of two method of operation: conduction and induction. *Conduction* is when power is transferred by physical means such as carbon brushes, slip rings, or commutators. *Induction* is when power is transferred by the principle of mutual inductance. The magnetic field of the field windings induces power into the rotor and causes it to turn.
Motors are powered by either an AC or a DC power source.

AC Motors

AC motors come in two types: *single-phase* and *three-phase* power. Whether to use a single-phase motor or a three-phase motor depends on what type of load is applied.

Three-Phase Motors   There are three basic types of three-phase motors:

1. Squirrel-cage induction motor
2. Wound-rotor induction motor
3. Synchronous motor

The main difference between these three is the type of rotor used. Rotors will be described in detail at each specific motor section later in this book. The squirrel-cage and wound-rotor motors are induction-type motors. The synchronous motor is a conduction motor. In earlier years, AC induction motors were referred to as rotating transformers because it utilizes the principle of mutual inductance. The stator winding is sometimes known as the primary winding, while the rotor is known as the secondary winding.

Single-Phase Motors   There are many different types of single-phase motors, and they have different operating principles.

1. Split-phase motors
2. Resistance-start induction-run motors
3. Capacitor-start induction-run motors
4. Capacitor-start capacitor-run motors
5. Shaded pole motors
6. Multispeed motors
7. Repulsion-type motors
8. Stepper motors
9. Universal motors

Single-phase motors have a wide variety of use, from home refrigerators, appliances, well pumps, and fans to being used in process control of a large industrial operation. Three-phase motors are predominately used for large commercial and industrial applications.

DC Motors

DC motors are made with one of three types of operational characteristics: series, shunt, and compound. Brushless, permanent magnet, and ServoDisc motors are compound type. DC motors are highly effective when variable speed is needed. The speed–torque characteristics of direct motors make them very useful when a small motor is needed.

DC motors can be configured in three different ways, depending on the type of speed control, torque, and horsepower required (see Figure 1–13).

Motor Construction

Electric motors have two main parts: the *stator,* sometimes called the *field winding,* and the *rotor,* sometimes called the *armature.* The terms *stator* and *field windings*—like *rotor* and *armature*—are used interchangeably, depending on the type of motor used. The stator and rotor produce magnetic fields that interact to create forces that turn the motor shaft. The stator, or field winding, is stationary. It does not turn. The stator is associated with AC motors. Other stationary parts include the frame, housings, covers, and bearings mounted in or on the motor as shown in Figure 1–14.

The magnetic field developed around the stator and rotor can be visualized as lines of force, or *flux.* The lines (flux or force) are in continuous loops from one magnetic pole to the other. When two magnets approach each other, the lines of force interact so that *like* magnetic poles (north-to-north and south-to-south) repel each other and *unlike* poles (north-to-south) attract each other.

The attraction and repulsion magnets are used to turn a shaft. If a rotor is placed crosswise between two stationary stators, the following occurs:

1. The attraction between like poles and the repulsion between unlike poles will produce a rotational force that will turn the rotor.
2. The rotor will continue to turn because the poles are changing between 50 and 60 cycles per second.

For more information on the theory of rotating magnetic fields see the "Theory and Calculations" volume of the *Electrician's Technical Reference.*

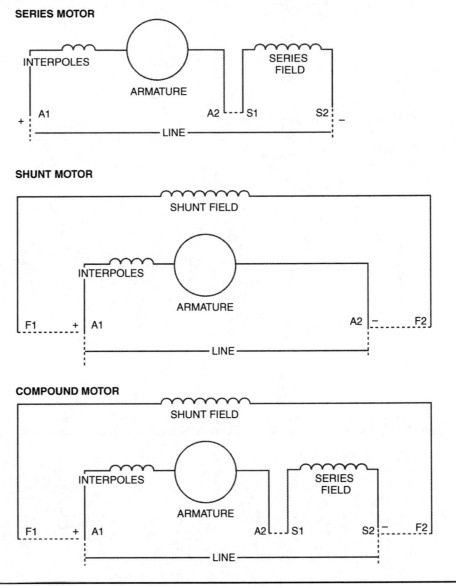

Figure 1–13 DC motor connections.

The stator core of a NEMA motor is made up of several hundred thin laminations (see Figure 1–15). Stator laminations are stacked together forming a hollow cylinder. Coils of insulated wire are inserted into slots of the stator core, as shown in Figure 1–16. Each grouping of coils, together with the steel core it surrounds, forms an electromagnet. Electromagnetism is the principle behind motor operation. An electromagnet is a coil of wire that carries current. The current produces a magnetic field just like the field around a permanent magnet. Usually the coil is wound on a steel core because steel carries the flux lines much better than air and allows a stronger field to be established with a given current. The stator windings are connected directly to the power source, as in Figure 1–17.

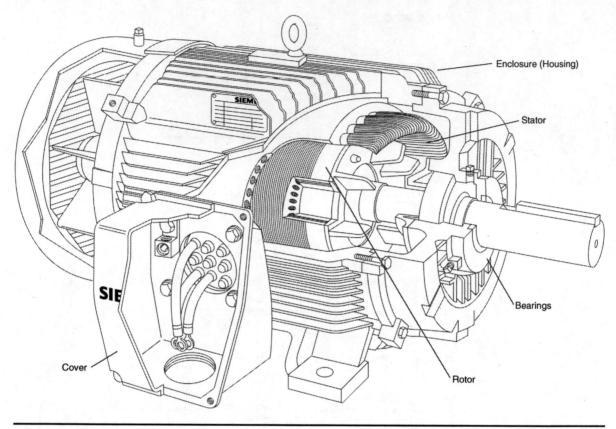

Enclosure (Housing)

Stator

Bearings

Rotor

Cover

Figure 1–14 AC motor construction; Courtesy Siemens Energy and Automation, Inc.

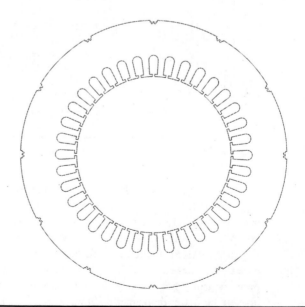

Figure 1–15 Stator construction; Courtesy Siemens Energy and Automation, Inc.

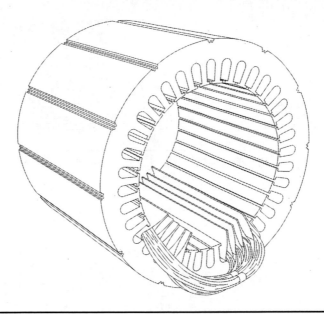

Figure 1–16   Stator windings; Courtesy Siemens Energy and Automation, Inc.

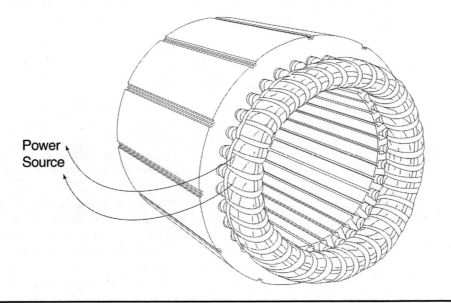

Figure 1–17   Stator winding connections; Courtesy Siemens Energy and Automation, Inc.

Rotor Construction

The rotor is the rotating part of the electromagnetic circuit. The most common type of rotor is the *squirrel-cage* rotor. Other types of rotor construction will be described later. The construction of the squirrel-cage rotor is reminiscent of rotating exercise wheels found in cages of pet rodents (see Figure 1–18).

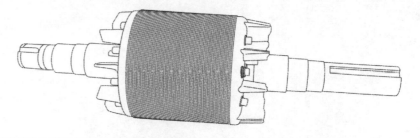

Figure 1–18   Rotating part (rotor); Courtesy Siemens Energy and Automation, Inc.

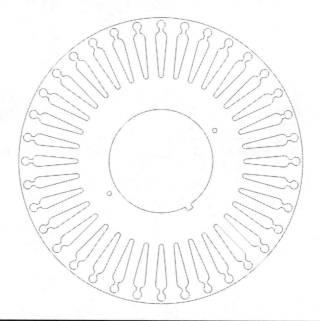

Figure 1–19   Rotor construction; Courtesy Siemens Energy and Automation, Inc.

The rotor consists of a stack of steel laminations with evenly spaced conductor bars around the circumference. See Figure 1–19. The laminations are stacked together to form a rotor core. Aluminum is die cast in the slots of the rotor core to form a series of conductors around the perimeter of the rotor. Current flow through the conductors forms the electromagnet. The conductor bars are mechanically and electrically connected with end rings. The rotor core mounts on a steel shaft to form a rotor assembly as shown in Figure 1–20.

Field Winding Construction

The pole pieces of salient-pole alternators may be built up of steel laminations, both as a manufacturing convenience and as a means of limiting the loss in their air-gap surfaces due to pulsations in air-gap flux. The field coils, wound directly on the poles or performed and then mounted on the poles, are suitably insulated from the poles for the voltages associated with normal and transient operation. The pole and coil assembly is bolted, dovetailed, or otherwise attached to the rotor body. The limitation of this attachment usually dictates when round-rotor construction must be used instead of salient-pole construction.

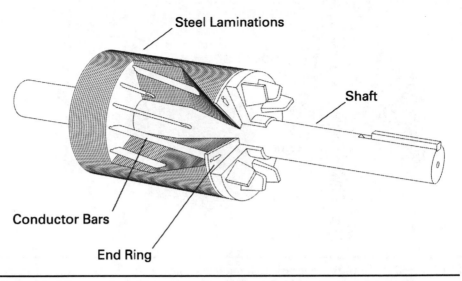

Steel Laminations

Shaft

Conductor Bars

End Ring

Figure 1–20   Conductor bars are mechanically and electrically connected with endrings; Courtesy Siemens Energy and Automation, Inc.

The rotor, or armature, is mounted on the motor shaft and turns with it. The rotor is associated with AC motors. Other rotating parts may include a commutator or a set of slip rings on the shaft.

Armature Construction

Armature cores are associated with DC motors. They are built up of thin laminations, produced as segments or rings, depending on size. Successive layers or groups of layers of the segmented laminations are staggered to minimize the effect of the joints in the magnetic circuit. The core is clamped between pressure plates and fingers to support it with sufficient pressure to prevent undue vibration of the laminations. Especially in long cores, the clamping arrangement may include some provision to compensate for compacting of the core after initial assembly.

The armature windings are fitted tightly in the slots and secured radially by *slot sticks,* or *wedges,* driven into suitable notches at the air-gap end of the slots. It is necessary that the stator coil ends be able to resist abnormal forces associated with short circuits. A supporting structure may be employed for this purpose. There are many variations of support design; most of them provide filler blocks between the coil sides—strategically located to transmit the circumferential forces from coil to coil—and additional structure to counteract the radial forces.

Coil supports ordinarily are designed to suit the need of a particular machine. Large two-pole machines require an elaborate structure; the combination of large short-circuit currents and coil ends—inherently flexible because of their long length—makes these machines particularly susceptible to coil-end movement. Low-speed machines with stiffer coil ends require less support; in the smallest ratings the coils may be capable of withstanding the short-circuit forces without any additional support.

Stator frames, or *yokes,* commonly are fabricated from structural steel, designed to support the core in proper alignment with the rotor and to suit the ventilating scheme used.

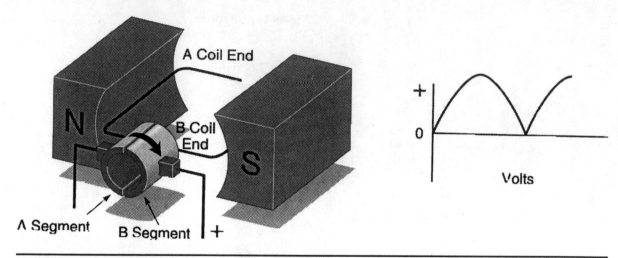

Figure 1–21 Brushes connected to commutator when coil is directly under the magnetic pole; Courtesy Herman, *Standard Textbook of Electricity,* Delmar Publishers.

The brushes ride on the slip rings or commutator to provide connection to the external circuit (see Figure 1–21).

Rotating Magnetic Fields

For continuous rotation to occur, the poles on either the rotor or stator must be reversed at the exact moment that they line up. This causes the poles, which had been attracting or repelling, to continue the attraction and repulsion action.

Current is reversed in the coils wound on the rotor in a DC motor. The stator in a DC motor, also has coils of wire, but current in them does not change. In some DC motors, the field poles are permanent magnets, though most use electromagnets.

In most AC motors, the powerlines are connected to the stator windings only. The stator poles reverse automatically as the current alternates. A few AC motors use a permanent magnet rotor, but most AC rotors have windings or a simple conducting cage of bars connected by end rings.

Stator Coil Arrangement

Figure 1–22 illustrates the relationship of the coils. In this example, six coils are used, two coils for each of the three phases. The coils operate in pairs. The coils are wrapped around the soft iron core material of the stator. These coils are referred to as *motor windings*. Each motor winding becomes a separate electromagnet. The coils are wound in such a way that when current flows in them, one coil is a north pole and its pair is a south pole. For example, if A1 were a north pole, then A2 would be a south pole. When current reverses direction the polarity of the poles would also reverse.

Developing a Rotating Magnetic Field

The principles of electromagnetism explain the shaft rotation of an AC motor. Recall that the stator of an AC motor is a hollow cylinder in which coils of insulated wire are inserted.

Figure 1–23 illustrates the relationship of the coils. In this example six coils are used, two coils for each of the three phases. The coils operate in pairs. The coils, or motor wind-

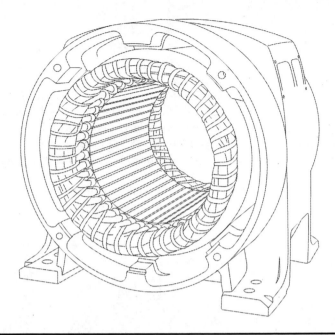

Figure 1–22 Shaft (rotor) turns when inserted in the hollow cylinder. The principle of electromagneticism explains how the shaft turns; Courtesy Siemens Energy and Automation, Inc.

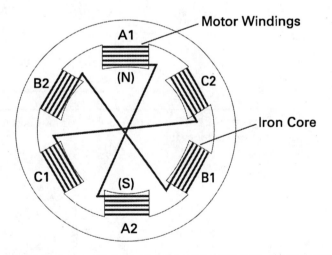

Figure 1–23 Stator coil arrangement; Courtesy Siemens Energy and Automation, Inc.

ings, are wrapped around the soft iron core material of the stator. Each motor winding becomes a separate electromagnet. The coils are wound in such a way that when current flows in them one coil is a north pole and its pair is a south pole. For example, if A1 were a north pole then A2 would be a south pole. When current reverses direction, the polarity of the poles would also reverse.

Power Supply

The stator is connected to a three-phase AC *power supply.* In Figure 1–24, phase A is connected to phase A of the power supply. Phases B and C would also be connected to phases B and C of the power supply, respectively.

Phase windings (A, B, and C) are placed electrical 120 degrees apart. In this example, a second set of three-phase windings is installed. The number of poles is determined by how many times a phase winding appears. In this example, each phase winding appears two times. This is a two-pole stator. If each phase winding appeared four times, it would be a four-pole stator. See Figure 1–25. Discuss the number of poles and degrees, electrical or mechanical.

When AC voltage is applied to the stator, current flows through the windings. The magnetic field developed in a phase winding depends on the direction of current flow through

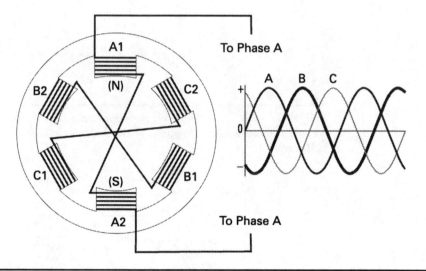

Figure 1–24   Connection to a three-phase power supply; Courtesy Siemens Energy and Automation, Inc.

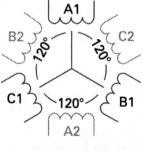

2-Pole Stator Winding

Figure 1–25   Number of poles is determined by how many times the phase winding appears; Courtesy Siemens Energy and Automation, Inc.

that winding. Table 1–6 will be used in the next few illustrations to demonstrate how a rotating magnetic field is developed. It assumes that a positive current flow in the A1, B1, and C1 windings result in a north pole.

Start Time

It is easier to visualize a magnetic field if a *start time* is picked when no current is flowing through one phase. In Figure 1–26, for example, a start time has been selected during which phase A has no current flow, phase B has current flow in a negative direction and phase C has current flow in a positive direction. Based on Table 1–6, B1 and C2 are south poles and B2 and C1 are north poles. Magnetic lines of flux leave the B2 north pole and enter the nearest south pole, C2. Magnetic lines of flux also leave the C1 north pole and enter the nearest south pole, B1. A magnetic field results, as indicated by the arrow.

TABLE 1.6   **Current Flow Direction Through the Winding**

Winding	Current Flow Direction	
	Positive	Negative
A1	North	South
A2	South	North
B1	North	South
B2	South	North
C1	North	South
C2	South	North

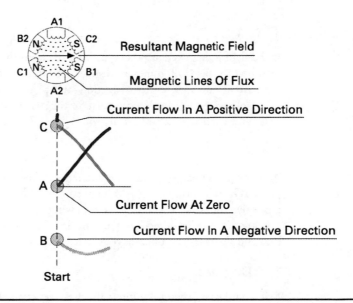

Figure 1–26 lines labeled: Resultant Magnetic Field, Magnetic Lines Of Flux, Current Flow In A Positive Direction, Current Flow At Zero, Current Flow In A Negative Direction, Start

Figure 1–26   Starting current flow and shaft rotation; Courtesy Siemens Energy and Automation, Inc.

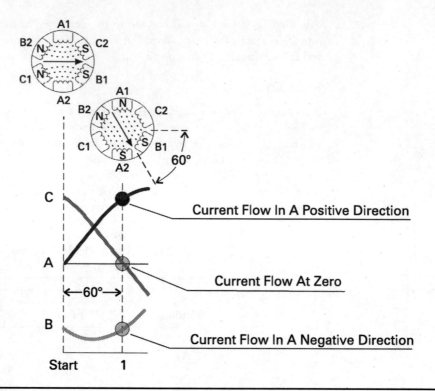

Figure 1–27 Current flow at 60° rotation, first time; Courtesy Siemens Energy and Automation, Inc.

Time 1

If the field is evaluated at 60 degree intervals from the starting point, at Time 1, it can be seen that the field will rotate 60 degrees. At Time 1, phase C has no current flow, phase A has current flow in a positive direction, and phase B has current flow in a negative direction. Following the same logic as used for the starting point, windings A1 and B2 are north poles and windings A2 and B1 are south poles. This is shown in Figure 1–27.

Time 2

At Time 2 the magnetic field has rotated another 60 degrees. As shown in Figure 1–28, phase B has no current flow. Although current is decreasing in phase A, it is still flowing in a positive direction. Phase C is now flowing in a negative direction. At start time it was flowing in a positive direction. Current flow has changed directions in the phase C windings and the magnetic poles have reversed polarity.

360 Degrees

At the end of six such time intervals, the magnetic field will have rotated one full revolution, or 360 degrees. This process will repeat 60 times a second on a 60-Hz power supply. See Figure 1–29.

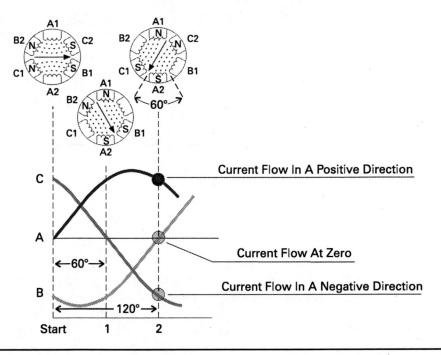

Figure 1–28 Current flow at 60 degrees, second time; Courtesy Siemens Energy and Automation, Inc.

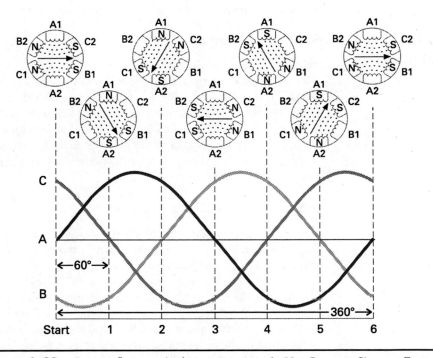

Figure 1–29 Current flow at 360 degrees rotation, 60 Hz; Courtesy Siemens Energy and Automation, Inc.

TABLE 1–7   Synchronous Speed at 60 Hz

Number of Poles	Synchronous Speed
2	3600
4	1800
6	1200
8	900
10	720

Synchronous Speed

The speed of the rotating magnetic field is referred to as *synchronous speed (Ns)*. Synchronous speed is equal to 120 times the frequency (*F*), divided by the number of poles (*P*).

$$Ns = \frac{120F}{P}$$

If the frequency of the applied power supply for the two-pole stator used in the previous example is 60 Hz, synchronous speed is 3600 RPM:

$$Ns = \frac{120 \times 60}{2}$$

$$Ns = 3600 \text{ RPM}$$

The synchronous speed decreases as the number of poles increase. Table 1–7 shows the synchronous speed at 60 Hz for the corresponding number of poles.

Rotor Rotation

Permanent Magnet

To see how a rotor works, a magnet mounted on a shaft can be substituted for the squirrel- cage rotor. When the stator windings are energized, a rotating magnetic field is established. The magnet's own magnetic field interacts with the rotating magnetic field of the stator. The north pole of the rotating magnetic field attracts the south pole of the magnet and the south pole of the rotating magnetic field attracts the north pole of the magnet. As the rotating magnetic field rotates, it pulls the magnet along, causing it to rotate. This type of design is used on some motors and is referred to as a *permanent magnet synchronous motor,* shown in Figure 1–30.

Induced Voltage Electromagnet

The squirrel-cage rotor acts essentially the same as the magnet. When power is applied to the stator, current flows through the winding, causing an expanding electromagnetic field that cuts across the rotor bars. See Figure 1–31.

When a conductor, such as a rotor bar, passes through a magnetic field, a voltage (emf) is induced in the conductor. The induced voltage causes a current flow in the conductor. Current flows through the rotor bars and around the end ring. The current flow in the conductor bars produces magnetic fields around each rotor bar. Recall that in an AC circuit current continuously changes direction and amplitude. The resultant magnetic field of the stator and rotor continuously change. The squirrel-cage rotor becomes an electromagnet with alternating north and south poles, as shown in Figure 1–32.

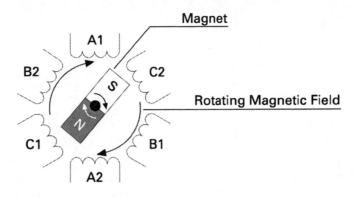

Figure 1–30 Permanent magnet motor; Courtesy Siemens Energy and Automation, Inc.

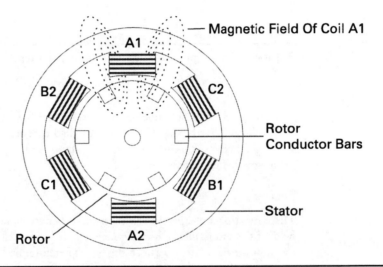

Figure 1–31 Induced voltage electromagnet; Courtesy Siemens Energy and Automation, Inc.

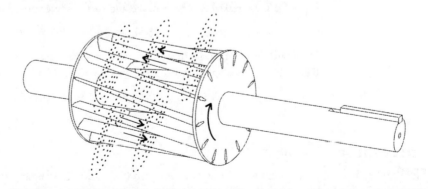

Figure 1–32 Rotor bar passing through a magnetic field; Courtesy Siemens Energy and Automation, Inc.

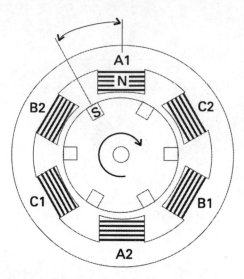

Figure 1–33 Opposite poles attract, like poles repel; Courtesy Siemens Energy and Automation, Inc.

Figure 1–33 illustrates one instant in time during which current flow through winding A1 produces a north pole. The expanding field cuts across an adjacent rotor bar, inducing a voltage. The resultant magnetic field in the rotor bar produces a south pole. As the stator magnetic field rotates, the rotor follows.

Slip

There must be a relative difference in speed between the rotor and the rotating magnetic field. If the rotor and the rotating magnetic field were turning at the same speed, no relative motion would exist between the two. Therefore, no lines of flux would be cut, and no voltage would be induced in the rotor. The difference in speed is called *slip*. Slip is necessary to produce torque. Slip is dependent on load. An increase in load will cause the rotor to slow down or increase slip. A decrease in load will cause the rotor to speed up or decrease slip. Slip is expressed as a percentage and can be determined with the following formula.

$$\% \text{ Slip} = Ns - Nr/Ns \times 100$$

For example, a four-pole motor operated at 60 Hz has a synchronous speed (Ns) of 1800 RPM. If the rotor speed at full load were 1750 RPM (Nr), the slip would be 2.8 percent.

$$\% \text{ Slip} = 1800 \times 100$$

$$\% \text{ Slip} = 1.9\%$$

Electric Motor Output

Horsepower

James Watt found that the average horse could do work at a rate of 550 ft lb/seconds. This became the basic horsepower measurement. Horsepower can also be expressed in the basic electrical unit for power, which is the watt.

$$1 \text{ horsepower} = 746 \text{ W}$$

Once horsepower has been converted to a basic unit of power, it can be converted to other power units, such as:

$$1 \text{ W} = 3.42 \text{ BTUs per hour}$$

$$1055 \text{ W} = 1 \text{ BTU per second}$$

$$4.19 \text{ W} = 1 \text{ calorie per second}$$

$$1.36 \text{ W} = 1 \text{ ft-lb. per second}$$

Horsepower is a measure of how fast and how much force of work is being done. In order to determine the horsepower output of a motor, the rate at which the motor is doing work must be known. The following formula determine the horsepower output of a motor:

$$HP = \frac{(1.59)\ (\text{torque})\ (\text{RPM})}{100,000}$$

Where:

HP = horsepower
1.59 = a constant
torque = torque in lbs/inch
RPM = speed; 100,000 = a constant

How to Determine the Horsepower of a Motor

Employ the following steps to determine horsepower using the input minus loss method for three-phase motors.

1. First measure the resistance of the motor coils per phase. Measure (T_1 to T_2 ohms) + (T_2 to T_3 ohms) + (T_1 to T_3 ohms). Divide by 6 to get ohms per phase.
2. Connect the motor to the line without a mechanical load. Measure the current average of the line leads. Use the current in the formula (12 × ohms per phase) × 3 to determine stator copper loss.
3. Measure the three-phase watts consumed by the motor at no load condition. The input watts minus the stator copper loss determines stray loss.
4. Load the motor to the load you wish to drive and measure the three-phase input watts again. Using the new input current, recalculate the stator copper loss now. Use loaded input watts minus loaded 12 × R of stator minus the stray losses constant to obtain rotor input watts.
5. Measure the speed of the rotor under load to calculate percentage of slip.

$$Ns - Nr \times 100 \div Ns = \% \text{ slip}$$

6. Rotor input watts × % slip = rotor copper loss.
7. Input watts – all losses (stray + stator + rotor) = output watts
8. Output watts ÷ 746 = output horsepower

Slip and Speed

The rotor must slip behind the synchronous speed in order to have voltage induced into it. In other words, the rotor bars must be cut by the magnetic field of the stator to induce voltage. Induced voltage creates rotor current that establishes the rotor magnetic poles, which are pushed and pulled by the stator poles. If the rotor traveled the same speed as the stator field, no current would be induced and no magnetic field would be set up to produce the twisting effort called *torque*.

The amount of slip depends on many factors: the design of the rotor, type of rotor conductors, air gap between the rotor and the stator iron, and mechanical load on the shaft. Slip is expressed in percentage and the formula is shown as follows. Slip is the difference between the rotating magnetic field's synchronous speed and the actual speed of the rotor. Slip is calculated as follows:

% Slip = Synchronous speed − Actual rotor speed ÷ Synchronous speed × 100

Speed Regulation

As the mechanical load on the motor shaft increases, the rotor slows down. This allows the rotor conductors to be cut at a higher rate of speed (the difference between synchronous speed and actual rotor speed). This effect of relative speed change induces more voltage in the rotor, creating more rotor current and increasing the strength of the rotor magnetic field. This enhances the magnetic pull and increases torque to compensate for the increased mechanical load.

The effect of changing load on the operating speed of the squirrel-cage rotor is reflected in the formula for percent speed regulation. This figure represents how well a motor will maintain its design speed over a wide range of load, from no load to full load. A perfect motor would have 0 percent speed regulation. This means there would be no speed variation between no load, and full-load speed. The formula for speed is

% Speed regulation = No load speed − Full load speed ÷ Full load speed × 100

This is not possible with an induction motor. There must be some change in speed to produce a change in current, which produces a change in torque to maintain a new, slower speed for a heavier load.

Torque

Magnetic lines of force flow in a direction of north to south between the poles of the stationary magnet. When magnetic lines of flux flow in the same direction, they repel each other.

When they flow in opposite directions, they attract each other. The magnetic lines of flux around the conductors cause the loop to be pushed the direction shown by the arrows. This pushing or turning force is called *torque* and is created by the magnetic field of the pole pieces and magnetic field of the loop or armature. Two factors determine the amount of torque produced by a direct current motor:

1. Strength of the magnetic field of the pole pieces
2. Strength of the magnetic field of the armature

Notice that there is no mention of speed or cutting action. One characteristic of a direct current motor is that it can develop maximum torque at 0 RPM.

The amount of torque produced by an AC induction motor is determined by three factors:

1. Strength of the magnetic field of the stator
2. Strength of the magnetic field of the rotor

3. Phase angle difference between rotor and stator fields

$$T = Kt \times \varphi \times Ir \times \cos \theta r$$

Where:

T = torque in ft lb

Kt = torque constant

φs = stator flux (constant at all speeds)

Ir = rotor current

$\cos \theta r$ = rotor power factor

Notice that one of the factors that determines the amount of torque produced by an induction motor is the strength of the magnetic field of the rotor. An induction motor can never reach synchronous speed. If rotors were to turn at the same speed as the rotating magnetic field, there would be no induced voltage in the rotor, and consequently, no rotor current. Without rotor current, there could be no magnetic field developed by the rotor and, therefore, no torque or turning force. A motor operating at no load will accelerate until the torque developed is proportional to the windage and bearing friction losses.

Torque, Voltage, and Current Relationship of the Motor Operation

If a load is connected to the motor, it must furnish more torque to operate the load. This causes the motor to slow down. When the motor speed decreases, the rotating magnetic field cuts the rotor bars at a faster rate. This causes more voltage to be induced in the rotor and, therefore, more current. The increased current flow produces a stronger magnetic field in the rotor, which causes more torque to be produced. The increased current flow in the rotor causes increased current flow in the stator. This is why *motor current will increase as load is added.*

Another factor that determines the amount of torque developed in an induction motor is the phase-angle difference between stator and rotor field flux. Maximum torque is developed when the stator and rotor flux are in phase with each other.

When a load is applied to the motor, it effects the voltage, the torque it exerts, its shaft speed, the horsepower it produces, the current, and the electrical power it draws from the distribution system. When there is no load on a motor, it exerts no shaft torque.

Motor Losses

The internal motor stray losses include eddy current losses in the rotor and stator and hysteresis losses in the iron. These are referred to as *iron losses* and actually are watts lost in heat; they convert electrical energy to mechanical energy. A second stray loss that is assumed to stay constant is *bearing friction.* This energy loss is consumed by the motor in the form of actual bearing friction. It takes some energy to keep the rotor moving against the bearing surfaces. Another stray loss is *windage loss.* Windage loss is the energy required to have the fan mounted on the motor to blow air over the motor and keep it cool. Most large motors have some type of fan that is designed to blow air over the motor housing. This fan is a drag on the motor and therefore consumes a small amount of energy.

The stray losses of iron loss, friction, and windage are assumed to stay constant from no load to full load. This is not true, but the losses do remain close enough in proportion to the remaining losses.

The other losses that occur in the motor are referred to as copper losses. These include the stator copper loss of $12 \times R$ of the stator winding and $12 \times R$ of the rotor winding.

These losses will change with load. The term *input minus loss* indicates that the input watts of the motor minus all the internal losses of the motor will yield the output of the motor in watts. Then use the watts-to-HP conversion (746W = 1 HP) to convert to mechanical horsepower.

Motor Efficiency

The efficiency of a motor is how much mechanical power it produces for the electrical power it consumes. It is expressed as a percentage of the input power. Efficiency of a motor depends on the design of the motor and the operating conditions. Motors tend to operate at maximum efficiency for the particular motor when operating at full load. This condition allows the maximum amount of transfer of energy from the stator to the rotor and losses in the motor make up a very small portion of the total input energy. Stray loss, friction, and windage loss are a relatively large portion of no-load energy. In fact, at no load the motor is 0 percent efficient because it does no work. However, as the motor is loaded, these losses remain fairly constant and are therefore a smaller part of the full-load energy.

Motor design also helps determine losses and efficiency. If the tolerance for the stator air gap (distance between rotor and stator) is very small, there is good magnetic transfer from rotor to stator and the motor is more efficient. If higher-quality bearings are used, friction is reduced and the motor becomes more efficient. If a better grade of silicon steel is used and laminations are thinner, the hysteresis and eddy current losses in the steel are reduced. Of course, if the motor is more efficient, it takes less energy to supply it for the same horsepower rating output of a less efficient motor. As with most advantages, there is a cost involved. High-efficiency motors cost more to purchase, but cost less to operate.

Watts input to the motor may be measured on a wattmeter. Remember that on AC motors you cannot simply multiply volts by amps to get watts. *Power factor (PF)* is another component with AC motors. To obtain true input watts (or true power) you must use $E \times I \times \%\ PF$ = watts, where E is voltage. Also to get three-phase input watts, check to be sure all the motor leads draw the same current and use the formula: $E \times I \times 1.73 \times \%\ PF$ = watts.

The other option for three-phase motors is use of a three-phase wattmeter or two single-phase wattmeters. They should be added according to Blondell's theorem.

The output horsepower is a mechanical measurement of work done. Convert this to electrical measurements to compare to electrical energy input. Multiply the horsepower output by 746 watts to get horsepower watts output. The formula to compute efficiency is: output watts/input watts $\times$ 100 = % efficiency.

Remember, motor efficiency is the percent of true power output compared to the true power input. See chapter 10 for the application of energy-efficient motors.

Motor Starting Current

Inrush current is known as the current flow at startup of a motor. If the rotor or armature is held during startup, a large current will be induced at the motor. When the motor is started it may be under load or unloaded. The startup current, then, is dependent on the condition in which the motor starts.

At startup the current will be close to that of the lock rotor current. A good rule of thumb to calculate startup current would be 75 percent of locked rotor current. However, when sizing overcurrent protection for a motor, it would be best to use the rule and tables given in the National Electrical Code. These rules and tables are further explained in chapter 2 of this book.

Locked rotor current is the current level when the rotational part (rotor or armature) of the motor is held when voltage is applied. It is sometimes called *first inrush current.*

Free rotor current is the current level at startup when the rotating part (rotor or armature) of the motor is not held and moving freely as motor starts. It is sometimes called the *second inrush current* because this current is approximately 75 percent of the peak or locked rotor value.

High inrush starting current of motors can cause voltage and power fluctuations within the premise wiring system or the power distribution system. This is one of the reasons for reduced voltage starting.

Power fluctuations can cause the following:

1. Operation of undervoltage devices

2. Stopping of motors at full load

3. Dips that cause problems with sensitive electronic equipment, lighting, and power circuits

4. Overloading of generators and alternators

Most power distribution companies have a policy concerning the starting currents of motors.

Motor Maintenance and Installation

Cleaning

A clean motor is more than just a pretty motor. Why is dirt bad? *Dirt* is a general word that can mean many things: dust, corrosive buildups, sugary syrups from food processing, electroconductive contaminants like salt deposits or coal dust, or even too many thick coats of paint. It can damage a motor in three ways. It can attack the electrical insulation by abrasion, or absorption into the insulation. It can contaminate lubricants and destroy bearings. It can obstruct heat removal. A clean motor runs cooler. Dirt builds up on the fan-cooled motor inlet openings and fan blades. This reduces the flow of air and increases the motor operating temperature. Dirt on the surface of the motor reduces heat transfer by convection and radiation. This is especially critical for totally enclosed motors, since all the cooling takes place on the outside surface. Heavily loaded motors are especially vulnerable to overheating, so they have little tolerance for dirt.

Surface dirt can be removed by various means, depending on its composition. Air pressure (30 psi maximum), vacuum cleaning. and direct wipe-down with rags or brushes are usually used. A recent development in cleaning is the use of dry ice (sand) blasting. This is usually done by a special contractor. The dry ice is less abrasive than mineral deposits, nonelectroconductive, and leaves no residue other than the removed dirt.

Dirt inside the motor is more difficult to remove. The best practice is to keep it from getting inside. A totally enclosed motor helps in this regard, but even an explosion-proof motor can be invaded and destroyed by fine dust. Some larger motors can be provided with a filter in the ventilation air circuit to keep dirt out. Keeping moisture out can reduce the attachment of dirt inside the motor and reduce the electrical conductivity of some contaminants. This will extend the time between disassembling the motor for cleaning.

Heat

Heat is a major concern to the effective operating of a motor. If the motor overheats its insulation can be damaged. Electrical power that is not converted into horsepower by a motor is converted into heat. Possible sources of heat are bearing friction, wind resistance, and core losses. Ambient temperature and internal temperature must be within the limits specified by the manufacturer.

Mountings, Couplings, and Alignment

Mounting is not normally a maintenance issue, but if it is inadequate, it can result in serious maintenance problems. The entire structure must be rigid, with a flat coplanar surface for the four mounting legs. The same applies to the structure for mounting the load. Both motor and load structure must be rigidly bound to the floor or a common structure. Failure to provide a solid mounting can lead to vibration or deflection with an ultimate consequence of bearing failure. Vertical motors can be even more demanding than horizontal motors because the mounting circle constitutes a small footprint for a large mass cantilevered above. Pliancy in the mounting structure can exacerbate low frequency vibration to which vertical motors are vulnerable. Always check hold-down bolts/dowels at every maintenance interval and do a visual check for cracks or other failures of the mounting system.

Coupling alignment is often promoted for energy efficiency. Energy loss in couplings is sometimes overstated, but proper alignment is always important to bearing and coupling life. Slight misalignment can dramatically increase the lateral load on bearings. It can also shorten the life of the coupling. One source attributes 45 to 80 percent of bearing and seal failures to misalignment.

Alignment means that the centerline of the motor and load shaft coincide. If they are parallel, but do not coincide, this is parallel misalignment. If the centerlines are not parallel but they intersect inside the coupling, this is angular misalignment. It is certainly possible to have misalignment in both respects.

Misalignment is usually the result of errors in installation. However, misalignment can sometimes develop after installation. This can occur if the mounting structure is not completely rigid, vibration or impact causes something to slip, or dirty or bent shims were originally used. Alignment should be checked soon after installation and less frequently after that if there are no conditions likely to cause misalignment. If there is evidence of misalignment, such as vibration, warm bearings or couplings, unusual noise, or rubber crumbs under the coupling, check alignment. Most users align couplings with a dial indicator. Mount the indicator on one shaft and contact the other coupling flange with the plunger parallel to the shaft to check for angular misalignment. Arrange the plunger radial to the flange to check for parallel misalignment. Rotate the shafts through at least 180 degrees and ideally 360 degrees, checking for runout. For misalignment in the horizontal plane, loosen motor-mounting bolts and reposition the motor. For any misalignment in the vertical plane, shims must be added or removed. Angular vertical misalignment requires unequal application of shims between shaft end and opposite end feet. Shims must be applied so that the motor rests with equal weight on both diagonal pairs of feet (i.e., there is no rocking of the motor before tightening bolts). Shims and the work area around motor feet must be completely clean. Discard any shims that are bent or scuffed. Always make a final alignment check after the mounting bolts are torqued.

Alignment is not easy. Many frustrating trial-and-error cycles of position and remeasure can be required. Computer programs can help in choosing shim size and positioning, but nothing makes it completely easy. In recent years, laser alignment devices have become the rage. They are more accurate and easier to use, especially where long extensions would be required to mount dial indicators. The laser devices simplify attachment, eliminate the problem of compliance of dial indicator mounting arms, and they are usually associated with (or directly connected to) computer devices that prescribe the adjustments necessary at all four legs.

Belt drives are entirely different from couplings, and they have their own needs for vigilant care. The most important thing is to control their tension. If belt drives are too loose, they tend to vibrate, wear rapidly, and waste energy though slippage. If they are too tight, they also will wear excessively, and they can dramatically shorten bearing life though excessive lateral loading.

Belt drives require parallel alignment between motor and load shaft, and require drive and driven pulleys to be in the same plane. Both of these conditions can usually be checked with a good straight edge. Once aligned, with a rigid base, alignment tends to hold constant much better than belt tension.

When belts appear worn or require over-tensioning to prevent slip, they should be replaced. Always replace multiple V-belts at the same time with a matched set. Recheck new belt tension several times until they complete break-in stretching (usually within the first 48 operating hours).

Consider replacing V-belts with synchronous belts (sometimes called cog belts) for elimination of slip loss. However, before doing so, determine whether slip is necessary in your application to protect the motor and load from jamming. Slip is sometimes necessary in systems that perform crushing or pumping of fluids with entrained solids. Operators sometimes rely upon the horrible screech to alert them that a jam has occurred.

Bearings

Some motors are designed for vertical mounts, with the bearings taking the weight of the motor and load without adverse wear. These motors' bearings are installed for heavy thrust loads, and are designed to carry loads that are applied parallel to the axis of the bearing. See Figure 1-34 for an example. Other motor bearings are designed to carry radial loads, as shown in Figure 1–35. Loads that act perpendicular to the axis of the motor shaft require radial bearings.

Radial loads can be supported by several types of bearings. Sleeve bearings can be used to support shafts. Generally, these bearings are cylindrical sleeves that slide over the shaft. They often use an oil reservoir or oil wick system in the housing to maintain the oil required for lubrication. If the bearings' lubrication fails or there is too much misalignment, the steel shaft will wear away the softer bronze bearing surface. To replace these bearings, remove the lubrication mechanism, then tap off the bronze sleeve.

Many motors use ball bearings to carry the radial load. Deep groove bearings frequently are used. Figure 1–36 shows various styles, including self-aligning bearings; double-row bearings for heavier, larger motors; and angular contact bearings that will support heavy thrust loads in one direction.

Other bearing styles include roller bearings. Roller bearings have higher load ratings than ball bearings of the same size and often are used in heavy-duty, lower-speed applications. See Figure 1–37.

As you check for bearing wear, you will need to listen and feel for suspected wear. Hot spots on the motor housing near the bearing race could indicate worn or failing bearings. Likewise, by listening to an operating motor, bad bearings can be heard as a grinding or gritty sound. After the motor is stopped and disconnected from power, hand turn the rotor.

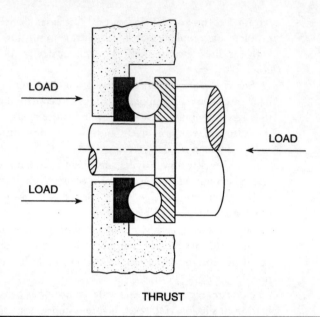

LOAD →

LOAD ←

LOAD →

THRUST

Figure 1–34 Thrust bearings are designed to carry leads that are applying pressure parallel to the shaft; Courtesy Keljik, *Electric Motors and Motor Controls,* Delmar Publishers.

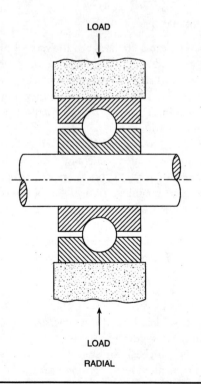

LOAD

LOAD

RADIAL

Figure 1–35 Radial bearings are bearings that carry a load that is perpendicular to the motor shaft; Courtesy Keljik, *Electric Motors and Motor Controls,* Delmar Publishers.

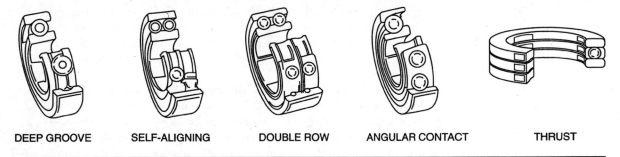

DEEP GROOVE   SELF-ALIGNING   DOUBLE ROW   ANGULAR CONTACT   THRUST

Figure 1–36 Various styles of ball bearings are used for specific applications; Courtesy Keljik, *Electric Motors and Motor Controls,* Delmar Publishers.

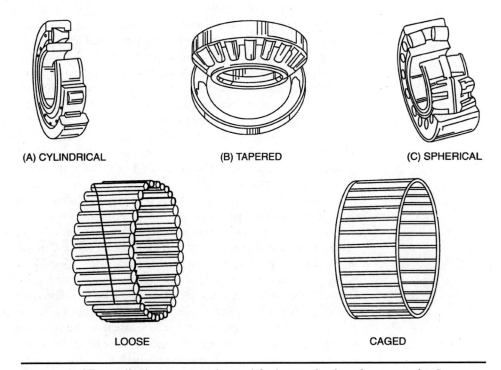

(A) CYLINDRICAL   (B) TAPERED   (C) SPHERICAL

LOOSE   CAGED

Figure 1–37 Roller bearings may be used for heavier loads or lower speeds; Courtesy Keljik, *Electric Motors and Motor Controls,* Delmar Publishers.

It should turn freely and easily. There should be imperceptible movement when moving the shaft in a radial direction and very little movement in the thrust direction. If there is excessive movement or unusual noise in either direction, further inspection may be necessary. If the bearings are worn or damaged, you will be able to measure the air gap around the circumference of the rotor. Use a feeler gauge to check the air gap between the rotor and the stator. Ideally, the top and bottom air gaps measurements should be identical.

Replace the bearings with similar types. Replacement with the exact bearing manufacturer is not necessary, as most bearing manufacturers adhere to consistent standards. Bearing pullers may be needed to remove the bearings. Use lubrication on the end of the cleaned shaft to pull the old bearing off the shaft. Pull on the inner race and avoid scarring the shaft.

To install the new bearings, reseat the bearing by driving the bearing inner race down the shaft's length. Be sure the race is aligned so as not to mar the shaft, and drive the bearing only with a soft metal tool against the inner race. Do not push against the outer race. Often the bearing can be heated slowly in an oven to expand the metal; it will then slip more easily over the shaft. As the bearing cools, it will contract around the shaft to create a firm bond. Similarly, the outer race should fit snugly into the bearing holder. The outside race should not move in its holder.

Sometimes bearings can be cleaned and relubricated to extend their lives. Bearings should be removed for most effective cleaning, but if that is not possible, cleaning and relubricating can take place on the shaft. Clean the bearing as much as possible with clean rags. To loosen old grease and hardened oil deposits, try using warmed kerosene. Be careful not to ignite the kerosene. Use low-pressure compressed air to blow the kerosene out of the bearing. Do not spin the ball bearings with the compressed air. Check the bearing for smooth operation. If there is no damage, the bearing can be relubricated and reinstalled.

Different bearings require different types of lubrication. Small, fractional horsepower motors often use sleeve bearings that are lubricated by oil. Spring-loaded caps cover the oil reservoir. Typically, these motors are oiled on an annual basis with an oil grade SAE 10–20. Larger sleeve-bearing motors use an oil well and a loose ring called a *slinger ring* that slings oil from the well up to a groove in the brass sleeve. Usually the oil becomes contaminated during the course of a year. The oil should be drained and refilled annually if the motor is used as a standard-duty motor.

Ball and roller bearings require lubrication as part of a preventive maintenance program. Check to be sure the bearings are not sealed bearings. Do not try to force grease into sealed bearings. If the bearings use grease as a lubrication method, the grease often can be applied using a grease gun. Clean the area around the grease inlet. (Often a zerk quick-attach grease fitting is used.) Remove the plug on the outlet side. If possible, run the motor. Add new grease to the inlet, forcing the old grease out and into a catch pan. Run the motor to allow the new grease to provide fresh lubrication to the motor. When no more old grease is forced out of the outlet hole, replace the outlet plug. Clean the motor and housing of excess grease. Be sure to use a grease that has adequate temperature ratings for the intended location of the motor. High-temperature greases may be necessary to keep the lubricant from separating and leaking out from the surrounding medium. Grease also helps provide better seals against grit and dirt than oil, if the proper grease is used for the application. **Remember to check bearings and ease of rotation regularly.**

Lubrication

Many small or integral horsepower motors have factory-sealed bearings that do not require relubrication. All others require lubrication. Unfortunately, lubrication can be more art than science. Motor manufacturers' recommendations should be followed initially. Eventually, with some experimentation and analysis of well-kept records, you may discover that a different type of lubricant or lubrication interval is better. It is good to compare experience with others in your same industry, because the operating environment greatly affects relubrication requirements. Consult with your motor repair shop. The repairer may be able to tell from inspecting your bearings and analyzing failures if you are using the wrong lubricants, lubrication methods, or intervals.

Typical lubrication intervals vary from less than three months for larger motors subjected to vibration, severe bearing loads, or high temperature to five years for integral horsepower motors with intermediate utilization. Motors used seasonally should be lubricated annually before the season of use.

One cannot merely be conservative and over-lubricate. There are many ways that improper lubrication shortens bearing life. Relubrication with a different grease can cause bearing failure when two incompatible greases mix. Grease consists of an oil in some type of constituent to give it body or thickness so that it doesn't run out of the bearing. Mixing greases with incompatible constituents can cause the components of the mixed grease to separate or harden. Table L is a guide to compatibility of grease bases.

Adding too much grease or greasing too frequently can force grease past the bearing shield or seal into the motor and damaged windings. Merely having too much grease in the bearing itself can prevent proper flow of the grease around the rollers. Sometimes bearing failures due to over-lubrication are interpreted as insufficient lubrication and intervals are made even shorter.

Perhaps the worst problem with greasing is introduction of contaminants when strict cleanliness standards are not followed in grease storage and application. It may be wise to buy grease in more expensive individual cartridges than large quantities that are subject to contamination when refilling grease guns. Take special care with grease fittings. Clean the fitting before filling and keep the grease gun nozzle covered when not injecting grease.

When selecting an oil or grease, again begin with manufacturers' recommendations. However, these sometimes are quite general or have allowed unexplained bearing failures. In this case, review alternative lubricant specifications and select a type compatible with the known contaminants in your operating environment. For severe situations, a synthetic lubricant may be best. Consult with lubricant vendors, the motor manufacturer, and repair shop regarding this. Lubricants vary in their tolerance to temperature, water, salt, or acids.

Finally, remember to completely remove old lubricant before trying a different one. If this is impossible, relubricate soon after introduction of a new lubricant. If there is a plug under a grease-lubricated bearing, remove this when first greasing with a new grease to encourage flushing of the old. Some authorities recommend running the motor for about an hour with the plug out to help flush the old grease. The Electrical Apparatus Service Association recommends removing the plug for all regreasing to purge the bearing of excess grease. The advisability of this depends on the geometry of the bearing cavities and type of seal/shield. Again, consult the manufacturer or repairer.

Maintenance Checks

Other maintenance checks can be performed. As mentioned, check for overheating and vibration. The motor housing and any air vents must be clean to allow cooling air to flow. Excess heat also may be caused by poor bearings.

Voltage and current measurements can be performed to be sure the motor is operating within the nameplate rating. Use a multimeter or a clamp-on meter. Use the voltmeter function to verify that the voltage delivered to the motor is within 10 percent of the motor's rated value. Use the clamp-on ammeter function to determine if the current draw is close to the nameplate rating, discussed at the beginning of this chapter. If the current is out of safe limits or if the current is unbalanced on a three-phase motor, investigate further.

Motor windings are insulated from each other by a film of varnish. Wire used in the motor winding that has a varnish insulation is called *magnet wire*. As a motor ages, the heating and cooling of the windings—as well as moisture, dirt, and oil—all work to break down the varnish-insulating qualities. The windings also are electrically insulated from the motor frame. Use of a megohmmeter will help you determine if there is a problem in the motor windings. (See Motor Testing section). The preventative maintenance checklist for motors at the end of this chapter will help you keep track of maintenance requirements.

Enclosures

Motor Enclosures

The enclosure consists of a *frame* (or yoke) and two *end brackets* (or bearing housings). The stator is mounted inside the frame. The rotor fits inside the stator with a slight air gap separating it from the stator. There is no direct physical connection between the rotor and the stator (see Figure 1–38).

The enclosure also protects the electrical and operating parts of the motor from harmful effects of the environment in which the motor operates. Bearings, mounted on the shaft, support the rotor and allow it to turn. A fan, also mounted on the shaft, is used on the motor shown in Figure 1–39 for cooling.

Motor Enclosure Types and Application

Many types of motor enclosures are available, including the ones discussed here.

Open An enclosure with ventilating openings that permit passage of external cooling air over and around the motor windings is open. This design is now seldom used.

Open Drip-Proof (ODP) An ODP is an open motor in which ventilation openings prevent liquids or solids from entering the machine at any angle less than 15 degrees from the

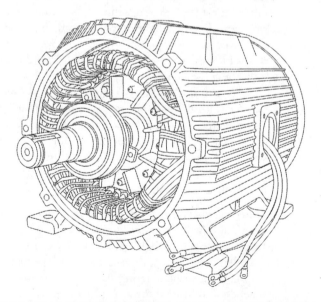

Figure 1–38 Enclosure (housing); Courtesy Siemens Energy and Automation, Inc.

vertical. Enclosure provides protection from contaminants in the environment in which the motor is operating. In addition, the type of enclosure affects the cooling of the motor. There are two categories of enclosures: open and totally enclosed.

Open enclosures permit cooling air to flow through the motor. The rotor has fan blades that assist in moving the air through the motor. One type of open enclosure is the drip-proof enclosure. The vent openings on this type of enclosure prevent liquids and solids falling from above at angles up to 15 degrees from vertical from entering the interior of the motor and damaging the operating components. When the motor is not in the horizontal position, such as mounted on a wall, a special cover may be necessary to protect it. This type of enclosure can be specified when the environment is free from contaminates, as shown in Figure 1–40.

Guarded   An open motor in which all ventilating openings are limited to specified size and shape is *guarded.*. This protects fingers or rods from accidental contact with rotating or electrical parts. Guarding can be accomplished by vaults, rooms, elevation or chain link fences.

Splash-Proof   A splash-proof motor is an open motor in which ventilation openings prevent liquid or solids from entering the machine at any angle less than 100 degrees from the vertical.

Totally Enclosed   A totally enclosed motor prevents the free exchange of air between the inside and outside of the case, but it is not airtight.

Totally Enclosed Nonventilated (TENV)   A TENV is a totally enclosed motor that is not equipped for cooling by means external to the exposed parts. In some cases, air surrounding the motor contains corrosive or harmful elements that can damage the internal parts of

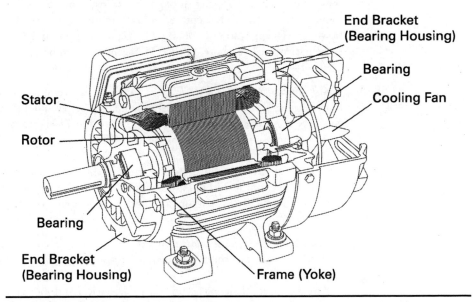

Figure 1–39   Electrical and operating parts; Courtesy Siemens Energy and Automation, Inc.

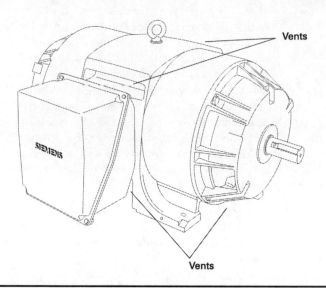

Figure 1–40 Open drip-proof (ODP); Courtesy Siemens Energy and Automation, Inc.

a motor. A totally enclosed motor enclosure restricts the free exchange of air between the inside of the motor and the outside. The enclosure is not airtight, however, and a seal at the point where the shaft passes through the housing keeps out water, dust, and other foreign matter that could enter the motor along the shaft. The absence of ventilating openings means all heat dissipates through the enclosure by means of conduction. Most TENV motors are fractional horsepower. TENV motors are used, however, for larger horsepower special applications. For larger horsepower applications, the frame is heavily ribbed to help dissipate heat more quickly. TENV motors can be used indoors and outdoors (see Figure 1–41).

Totally Enclosed Fan-Cooled (TEFC) A TEFC is a totally enclosed motor with a fan to blow cooling air across the external frame. They are commonly used in dusty, dirty, and corrosive atmospheres. The TEFC is similar to the TENV except that an external fan is mounted opposite the drive end of the motor. The fan provides additional cooling by blowing air over the exterior of the motor to dissipate heat more quickly. A shroud covers the fan to prevent anyone from touching it. With this arrangement, no outside air enters the interior of the motor. TEFC motors can be used in dirty, moist, or mildly corrosive operating conditions. TEFC motors are more widely used for integral HP applications (see Figure 1–42).

Encapsulated An encapsulated motor is an open motor in which the windings are covered with a heavy coating of material to provide protection from moisture, dirt, and abrasion.

Enclosures are designed to protect the electrical and mechanical equipment from different types of environment. *NEMA* (National Electrical Manufacturers Association) has a design standard for different applications. A partial list and general information are given

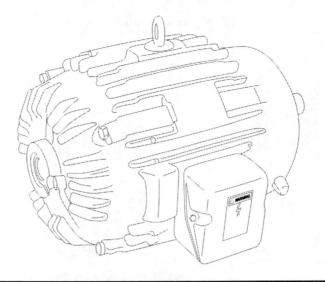

Figure 1–41 Totally enclosed non-ventilated (TENV) motor; Courtesy Siemens Energy and Automation, Inc.

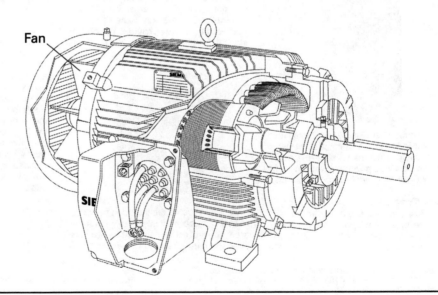

Figure 1–42 Totally enclosed fan cooled (TEFC) motor; Courtesy Siemens Energy and Automation, Inc.

in Figure 1–43. *IEC* (International European Commission) also has ratings of enclosures, given in Figure 1–44.

Explosion Proof (XP) The explosion-proof motor enclosure is similar in appearance to the TEFC; however, most XP enclosures are cast iron. The application of motors used in hazardous locations is subject to regulations and standards set by regulatory agencies such

**NEMA Type 1
General Purpose
Surface Mounting**

Type 1 enclosures are intended for indoor use primarily to provide a degree of protection against contact with the enclosed equipment in locations where unusual service conditions do not exist. The enclosures are designed to meet the rod entry and rust-resistance design tests. *Enclosure is sheet steel, treated to resist corrosion.*

**NEMA Type 1
Flush Mounting**

Flush mounted enclosures for installation in machine frames and plaster wall. These enclosures are for similar applications and are designed to meet the same tests as NEMA Type 1 surface mounting.

NEMA Type 3
Type 3 enclosures are intended for outdoor use primarily to provide a degree of protection against windblown dust rain and sleet; and to be undamaged by the formation of ice on the enclosure They are designed to meet rain ■, external icing ■, dust and rust-resistance design tests. They are not intended to provide protection against conditions such as internal condensation or internal icing.

NEMA Type 3R

Type 3R enclosures are intended for outdoor use primarily to provide a degree of protection against falling rain, and to be undamaged by the formation of ice on the enclosure. They are designed to meet rod entry, rain ■, external icing ■, and rust-resistance design tests. They are not intended to provide protection against conditions such as dust internal condensation, or internal icing.

NEMA Type 4

Type 4 enclosures are intended for indoor or outdoor use primarily to provide a degree of protection against windblown dust and rain, splashing water, and hosedirected water; and to be undamaged by the formation of ice on the enclosure They are designed to meet hosedown dust, external icing ■, and rust-resistance design tests They are not intended to provide protection against conditions such as internal condensation or internal icing. Enclosures are made of heavy gauge stainless steel, cast aluminum or heavy gauge sheet steel, depending on the type of unit and size. Cover has a synthetic rubber gasket.

**NEMA Type 3R, 7 & 9
Unilock Enclosure
For Hazardous
Locations**

This enclosure is cast from "copper-free" (less than 0.1%) aluminum and the entire enclosure (including interior and flange areas) is bronze chromated The exterior surfaces are also primed with a special epoxy primer and finished with an aliphatic urethane paint for extra corrosion resistance. The V-Band permits easy removal of the cover for inspection and for making field modifications. This enclosure meets the same tests as separate NEMA Type 3R, and NEMA Type 7 and 9 enclosures. For NEMA Type 3R application, it is necessary that a drain be added

■ Evaluation criteria: No water has entered enclosure during specified test.
■ Evaluation criteria: Undamaged after ice which built up during specified test has melted (Note: **Not** required to be operable while ice-laden).
■ Evaluation criteria: No water shall have reached live parts, insulation or mechanisms.

Figure 1–43 NEMA descriptions and examples in metal and fiberglass; Courtesy of Rockwell Automation/ Allen-Bradley.

**NEMA Type 4X
Non-Metallic,
Corrosion-Resistant
Fiberglass Reinforced
Polyester**

Type 4X enclosures are intended for indoor or outdoor use primarily to provide a degree of protection against corrosion, windblown dust and rain, splashing water, and hose-directed water; and to be undamaged by the formation of ice on the enclosure. They are designed to meet the hosedown, dust, external icing, and corrosion-resistance design tests. They are not intended to provide protection against conditions such as internal condensation or internal icing. Enclosure is fiberglass reinforced polyester with a synthetic rubber gasket between cover and base. Ideal for such industries as chemical plants and paper mills.

NEMA Type 6P

Type 6P enclosures are intended for indoor or outdoor use primarily to provide a degree of protection against the entry of water during prolonged submersion at a limited depth; and to be undamaged by the formation of ice on the enclosure. They are designed to meet air pressure, external icing, hosedown and corrosion-resistance design tests. They are not intended to provide protection against conditions such as internal condensation or internal icing.

**NEMA Type 7
For Hazardous
Gas Locations
Bolted Enclosure**

Type 7 enclosures are for indoor use in locations classified as Class I, Groups C or D, as defined in the National Electrical Code. Type 7 enclosures are designed to be capable of withstanding the pressures resulting from an internal explosion of specified gases, and contain such an explosion sufficiently that an explosive gas-air mixture existing in the atmosphere surrounding the enclosure will not be ignited. Enclosed heat generating devices are designed not to cause external surfaces to reach temperatures capable of igniting explosive gas-air mixtures in the surrounding atmosphere. Enclosures are designed to meet explosion, hydrostatic, and temperature design tests. Finish is a special corrosion-resistant, gray enamel.

**NEMA Type 9
For Hazardous
Dust Locations**

Type 9 enclosures are intended for indoor use in locations classified as Class II, Groups E, F or G, as defined in the National Electrical Code. Type enclosures are designed to be capable of preventing the entrance of dust. Enclosed heat generating devices are designed not to cause external surfaces to reach temperatures capable of igniting or discoloring dust on the enclosure or igniting dust-air mixtures in the surrounding atmosphere. Enclosures are designed to meet dust penetration and temperature design tests, and aging of gaskets. The outside finish is a special corrosion-resistant gray enamel.

NEMA Type 12

Type 12 enclosures are intended for indoor use primarily to provide a degree of protection against dust, falling dirt, and dripping non-corrosive liquids. They are designed to meet drip, dust, and rust-resistance tests. They are not intended to provide protection against conditions such as internal condensation.

NEMA Type 13

Type 13 enclosures are intended for indoor use primarily to provide a degree of protection against dust, spraying of water, oil, and non-corrosive coolant. They are designed to meet oil exclusion and rust-resistance design tests. They are not intended to provide protection against conditions such as internal condensation.

Evaluation criteria: No water has entered enclosure during specified test
Evaluation criteria: Undamaged after ice which built up during specified test has melted (Note: **Not** required to be operable while ice-laden).

ENCLOSURES
IEC

G E N E R A L

DEGREE OF PROTECTION

IEC Publication 529 describes standard Degrees of Protection which enclosures of a product are designed to provide when properly installed.

SUMMARY

The publication defines degrees of protection with respect to:
● Persons
● Equipment within the enclosure
● Ingress of water

It does **not** define:
● Protection against risk of explosion
● Environmental protection (e.g. against humidity, corrosive atmospheres or fluids, fungus or the ingress of vermin)

NOTE: The IEC test requirements for Degrees of Protection against liquid ingress refer only to water. Those products in this catalog, which have a high degree of protection against ingress of liquid, in most cases include Nitrile seals. These have good resistance to a wide range of oils, coolants and cutting fluids. However, some of the available lubricants, hydraulic fluids and solvents can cause severe deterioration of Nitrile and other polymers. Some of the products listed are available with seals of Viton or other materials for improved resistance to such liquids. For specific advice on this subject refer to your nearest Allen-Bradley Sales Office.

IEC ENCLOSURE CLASSIFICATION

The degree of protection is indicated by two letters (IP) and two numerals. International Standard IEC 529 contains descriptions and associated test requirements which define the degree of protection each numeral specifies. The following table indicates the *general* degree of protection — refer to Abridged Descriptions of IEC Enclosure Test Requirements below and on Page 193. **For complete test requirements refer to IEC 529.**

FIRST NUMERAL■	SECOND NUMERAL■
Protection of persons against access to hazardous parts and protection against penetration of solid foreign objects.	Protection against ingress of water under test conditions specified in IEC 529.
0 Non-protected	**0** Non-protected
1 Back of hand; objects greater than 50mm in diameter	**1** Vertically falling drops of water
2 Finger; objects greater than 12.5mm in diameter	**2** Vertically falling drops of water with enclosure tilted 15 degrees
3 Tools or objects greater than 2.5mm in diameter	**3** Spraying water
4 Tools or objects greater than 1.0mm in diameter	**4** Splashing water
5 Dust-protected (Dust may enter during specified test but must not interfere with operation of the equipment or impair safety)	**5** Water jets
6 Dust tight (No dust observable inside enclosure at end of test)	**6** Powerful water jets
	7 Temporary submersion
	8 Continuous submersion

Example: IP41 describes an enclosure which is designed to protect against the entry of tools or objects greater than 1mm in diameter and to protect against vertically dripping water under specified test conditions.

Note: All first numerals, and second numerals up to and including characteristic numeral **6**, imply compliance also with the requirements for all lower characteristic numerals in their respective series (first or second). Second numerals **7** and **8** do **not** imply suitability for exposure to water jets (second characteristic numeral **5** or **6**) unless dual coded; e.g. **IP_5/IP_7.**

■ The IEC standard permits use of certain supplementary letters with the characteristic numerals. If such letters are used, refer to IEC 529 for the explanation.

Figure 1–44   IEC designations as I P numbers. Three NEMA to IEC comparisons; Courtesy of Rockwell Automation/Allen-Bradley.

as the National Electrical Code (NEC) and Underwriters Laboratories (UL) for XP motors used in the United States (see Figure 1–45).

Hazardous Environments

Although you should never specify or suggest the type of location, it is important to understand regulations that apply to hazardous locations. It is the user's responsibility to contact local regulatory agencies to define the location as either Division I or II and to comply with all applicable codes.

Hazardous materials are normally present in the atmosphere. A division I location requires an explosion-proof motor.

Atmosphere may become hazardous as result of abnormal conditions. This may occur if, for example, a pipe breaks that is the conduit for a hazardous chemical.

ENCLOSURES
IEC

ABRIDGED DESCRIPTIONS OF IEC ENCLOSURE TEST REQUIREMENTS — Cont'd

Tests for protection against access to hazardous parts (first characteristic numeral) — Cont'd

IP4_ — A test wire 1mm in diameter shall not penetrate and adequate clearance shall be kept from hazardous live parts (as specified on Page 18). Force = 1 N.

IP5_ — A test wire 1mm in diameter shall not penetrate and adequate clearance shall be kept from hazardous live parts (as specified on Page 18). Force = 1 N.

IP6_ — A test wire 1mm in diameter shall not penetrate and adequate clearance shall be kept from hazardous live parts (as specified on Page 18). Force = 1 N.

Tests for protection against solid foreign objects (first characteristic numeral)

For first numerals **1, 2, 3** and **4** the protection against solid foreign objects is satisfactory if the full diameter of the specified probe does not pass through any opening. Note that for first numerals **3** and **4** the probes are intended to simulate foreign objects which may be spherical. Where shape of the entry path leaves any doubt about ingress of a spherical object capable of motion, it may be necessary to examine drawings or to provide special access for the object probe. For first numerals **5** and **6** see test descriptions below for acceptance criteria.

IP0_ — No test required.

IP1_ — The full diameter of a rigid sphere 50mm in diameter must not pass through any opening at a test force of 50 N.

IP2_ — The full diameter of a rigid sphere 12.5mm in diameter must not pass through any opening at a test force of 30 N.

IP3_ — A rigid steel rod 2.5mm in diameter must not pass through any opening at a test force of 3 N.

IP4_ — A rigid steel wire 1mm in diameter must not pass through any opening at a test force of 1 N.

IP5_ — The test specimen is supported inside a specified dust chamber where talcum powder able to pass through a square-meshed sieve with wire diameter 50 µm and width between wires 75 µm, is kept in suspension.

Enclosures for equipment subject to thermal cycling effects (category 1) are vacuum pumped to a reduced internal pressure relative to the surrounding atmosphere: maximum depression = 2 kPa; maximum extraction rate = 60 volumes per hour. If extraction rate of 40 to 60 volumes/hr. is obtained, test is continued until 80 volumes have been drawn through or 8 hr. has elapsed. If extraction rate is less than 40 volumes/hr. at 20 kPa depression, test time = 8 hr.

Enclosures for equipment not subject to thermal cycling effects **and** designated category 2 in the relevant product standard are tested for 8 hr. without vacuum pumping.

Protection is satisfactory if talcum powder has not accumulated in a quantity or location such that, as with any other kind of dust, it could interfere with the correct operation of the equipment or impair safety; and no dust has been deposited where it could lead to tracking along creepage distances.

IP6_ — All enclosures are tested as category 1, as specified above for **IP5_**. The protection is satisfactory if no deposit of dust is observable inside the enclosure at the end of the test.

Tests for protection against water (second characteristic numeral)

The second characteristic numeral of the IP number indicates compliance with the following tests for the degree of protection against water. For numerals **1** through **7**, the protection is satisfactory if any water which has entered does not interfere with satisfactory operation, does not reach live parts not designed to operate when wet, and does not accumulate near a cable entry or enter the cable. For second numeral **8** the protection is satisfactory if no water has entered the enclosure.

IP_0 — No test required.

IP_1 — Water is dripped onto the enclosure from a "drip box" having spouts spaced on a 20mm square pattern, at a "rainfall" rate of 1 mm/min. The enclosure is placed in its normal operating position under the drip box. Test time = 10 min.

IP_2 — Water is dripped onto the enclosure from a "drip box" having spouts spaced on a 20mm square pattern, at a "rainfall" rate of 3 mm/min. The enclosure is placed in 4 fixed positions tilted 15° from its normal operating position, under the drip box. Test time = 2.5 min. for each position of tilt.

IP_3 — Water is sprayed onto all sides of the enclosure over an arc of 60° from vertical, using an oscillating tube device with spray holes 50mm apart (or a hand-held nozzle for larger enclosures). Flow rate, oscillating tube device = 0.07 l/min. per hole x number of holes; for hand-held nozzle = 10 l/min. Test time, oscillating tube = 10 min.; for hand-held nozzle = 1 min./m² of enclosure surface area, 5 min. minimum.

IP_4 — Same as test for **IP_3** except spray covers an arc of 180° from vertical.

IP_5 — Enclosure is sprayed from all practicable directions with a stream of water at 12.5 l/min. from a 6.3mm nozzle from a distance of 2.5 to 3m. Test time = 1 min./m² of enclosure surface area to be sprayed, 3 min. minimum.

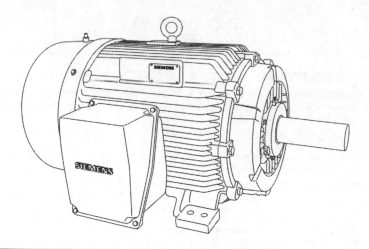

Figure 1–45   Explosion-proof (XP); Courtesy Siemens Energy and Automation, Inc.

TABLE 1–8   Classes of Hazardous Materials

Class I	Class II	Class III
Divisions I & II	Divisions I & II	Divisions I & II
Groups A–D	Groups E–G	No Groups
Gases & vapors	Flammable Dust	Ignitable Fibers

Once the location is defined as hazardous, the location is further defined by the class and group of hazard (see Table 1–8). Class I, Groups A through D, are chemical gases or liquids such as gasoline, acetone, and hydrogen. Class II, Groups E, F, and G, include flammable dust, such as coke or grain dust. Class III is not divided into groups. It includes all ignitable fiber lints, such as cloth in a fiber in textile mills.

In some cases it may be necessary for the user to define the lowest possible ignition temperature of the hazardous material to assure the motor complies with all applicable codes and requirements in the future. See section "Hazardous Locations" and the NEC, 1996 edition, articles 500–504.

Enclosures for Above NEMA Motors

Open Drip-Proof (ODP)   Environmental factors also affect large AC motors. Enclosures used on above NEMA motors look different from those on integral-frame-size motors. The open drip-proof enclosure provides the same amount of protection as the integral-frame-size open motor. This provides the least amount of protection for the motor's electrical components. It is typically used in environments free of contaminants (see Figure 1–46).

Horizontal Drip-Proof Weather-Protected I (Type CG) and Weather-Protected I (Type CGII)   The weather-protected I enclosure is an open enclosure that has ventilating passages designed to minimize the entrance of rain, snow, and airborne particles that could come into contact with the electrical and rotating parts of the motor. All air inlets and exhaust vents are covered with screens. It is used on indoor applications when there is a small amount of moisture in the air. This enclosure is available to 10,000 horsepower (see Figure 1–47).

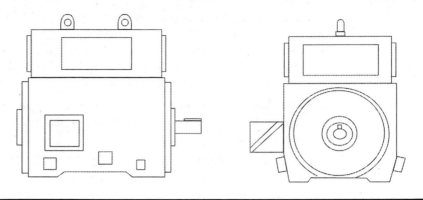

Figure 1–46   Open drip-proof; Courtesy Siemens Energy and Automation, Inc.

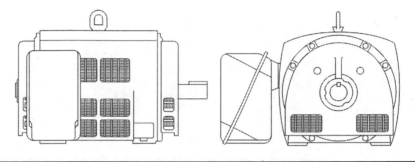

Figure 1–47   Horizontal drip-proof weather-protected I (type CG); Courtesy Siemens Energy and Automation, Inc.

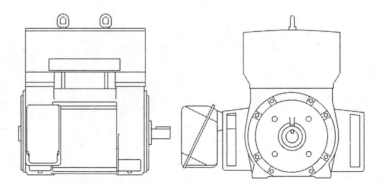

Figure 1–48   Weather-protected II (type CGII); Courtesy Siemens Energy and Automation, Inc.

Weather-Protected II (Type CGII)   Weather-protected II enclosures are open enclosures with vents constructed so that high-velocity air and airborne particles blown into the motor can be discharged without entering the internal ventilating passages leading to the electrical parts of the motor. The intake and discharge vents must have at least three 90-degree turns and the air velocity must be less than 600 feet per minute. It is used outdoors when the motor is not protected by other structures. This enclosure is available through 10,000 horsepower (see Figure 1–48).

Totally Enclosed Fan-Cooled (Type CGZ)   The totally enclosed fan-cooled motor functions the same as the TEFC enclosure used on integral-frame-size motors. It is designed for indoor and outdoor applications where internal parts must be protected from adverse ambient conditions. Type CGZ uses cooling fins all around the yoke and housing, and is available up to 900 HP on 580 frames and 2250 HP on 708–880 frames. See Figure 1–49.

Totally Enclosed Air-to-Air Cooled (Type TEAAC)   Type TEAAC totally enclosed motor utilizes air-to-tube-type heat exchangers for cooling. It is available through 4500 HP. See Figure 1–50.

Totally Enclosed Water-to-Air Cooled (Type TEACC)   There comes a point when the motor frame cannot adequately dissipate heat, even with the help of a fan. This enclosure is designed to cool the motor by means of a water-to-air heat exchanger. This type of enclosure requires a steady supply of water. It is available through 10,000 HP. See Figure 1–51.

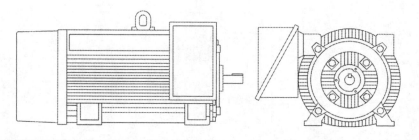

Figure 1–49   Totally enclosed fan-cooled (typs CGII); Courtesy Siemens Energy and Automation, Inc.

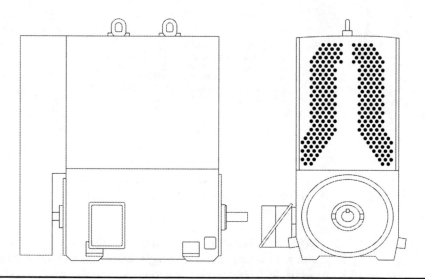

Figure 1–50   Totally enclosed air-to-air cooled (type TEAAC); Courtesy Siemens Energy and Automation, Inc.

Totally Enclosed Fan-Cooled Explosion-Proof (Type AZZ)   Large AC motors are also used in hazardous environments. This enclosure meets or exceeds all applicable UL requirements for hazardous (Division I) environmental operation. It is available through 1750 HP. See Figure 1–52.

Brake

Their are many ways to brake a motor. To brake a motor we must stop the rotating member of the motor. A brake is an electromechanical device installed on the shaft of the motor. There are four typical methods of braking:

- **Dynamic braking.** This method uses the CEMF produced by the motor. This method is accomplished by how circuitry is applied to the motor. This is not an external mechanical device such as mechanical or disk brakes. For more information on how this is accomplished see braking control section in the motor control volume.

- **DC braking.** This method uses the principle that the rotor will try to follow the stator field. Applying DC to the stator field of an AC motor after AC is removed

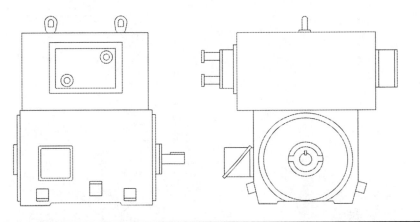

Figure 1–51   Totally enclosed water-to-air cooled (type RGG); Courtesy Siemens Energy and Automation, Inc.

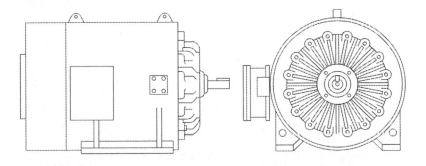

Figure 1–52   Totally enclosed fan-cooled explosion-proof (type A22); Courtesy Siemens Energy and Automation, Inc.

will hold the stator in place. This braking is accomplished by electrical circuitry, not by mechanical means such mechanical and disk brakes. For more information on how this is accomplished, see the braking control section in the motor control volume.

- **Mechanical brakes.** This method is accomplished with a mechanical drum and shoe brakes used to keep the motor shaft from moving. The brake drum is attached to the motor shaft and the brake shoes are used to hold the drum in place. In most applications, the brake is clamped in the closed position by a heavy spring when there is no power on the motor or brake.

This is a fail-safe system that applies the brake in case of an electrical failure. When the power is applied to the motor, the brake solenoid coil is also energized and pulls the brake shoes away from the drum. Figure 1–53 shows how the brake solenoid is connected into the electrical circuit.

- **Disk brakes.** Two different methods are used to apply this type braking action. One style shows the brake disk and the brake pads mounted on the side. As the disk

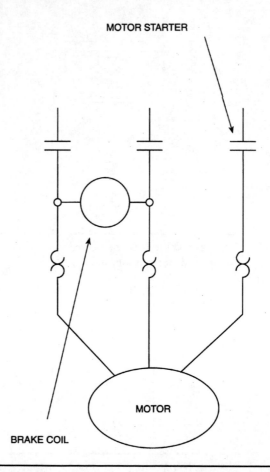

Figure 1–53 Electrical brakes are connected on the load side of the motor starter. It is best to connect them ahead of the motor overloads; Courtesy Keljik, *Electric Motors and Motor Controls,* Delmar Publishers.

rotates, the pads are pulled off via electrical solenoids. As the power to the motor is interrupted, the pads will be pressed against the rotating disk to stop the motor. Some disks are applied to the rotor by electromagnetic action. This means that the braking action can be controlled. In some models, the higher the DC current to the brake, the harder the pads are pressed against the disks. In other styles, a fail-safe system is used. *Fail-safe* means that permanent magnets are trying to force the pads to maximum contact pressure. Electric current must be applied to the brake to pull the pads away from the disk. With the fail-safe system, maximum braking will occur if there is an electrical power failure or failure of the DC brake supply voltage.

Preventive Motor Maintenance Checklist

The following are helpful maintenance checks that could prolong the life of your motor. A complete list of the motor operation, functions, and past history will help determine the frequency of inspection needed.

- **Stator and rotor windings.** The life of a winding depends on keeping it as near to its original condition as long as possible. Insulation failure causes immediate outage time. The following points should be carefully examined and corrective action taken during scheduled inspections to prevent operational failures.

  1. **Clean dust and dirt.** Dust and dirt are almost always present in windings that have been in operation under average conditions. Some forms of dust are highly conductive and contribute materially to insulation breakdown, as well as restricting ventilation. Note evidence of moisture, oil, or grease on the winding and, if necessary, clean the winding thoroughly with a solvent solution. Generally, after a major cleaning, a drying process is required to restore the insulation to a safe level for operation.

  2. **Check winding tightness** in the slots or on the pole pieces. One condition that hastens winding failure is movement of the coils due to vibration during operation. The effects of varnish and oven treatments so as to fill all air spaces caused by drying and shrinkage of the insulation will maintain a solid winding.

  3. **Check insulation surfaces** for cracks, crazing, flaking, powdering, or other evidence of need to renew insulation. Usually under these conditions, when the winding is still tight in the slots, a coat or two of air-drying varnish will restore the insulation to a safe value.

  4. **Check the winding mechanical supports** for insulation quality and tightness, the ring binding on stator windings, and the glass or wire-wound bands on rotating windings.

  5. **Examine squirrel-cage rotors** for excessive heating or discolored rotor bars that may indicate open circuits or high resistance points between the end rings and rotor bars. The symptoms of such conditions are slowing down under load and reduced starting torque. Brazing broken bars or replacing bars should be done only by a competent person or repair shop.

  6. **Check brushes, collector rings, and commutators.** In general, observe the machine in operation if possible and note any evidence of missoperation such as sparking, chatter of brushes in the holder, or lack of cleanliness as an aid to inspection repairs later.

7. **Select proper brushes.** Successful brush operation depends on the proper selection of the brush most suitable for the service requirements.

8. **Check brushes in holders** for fit and free play and replace those that are worn down almost to the brush rivet.

9. **Tighten brush studs** that may have become loose from the drying and shrinking of insulating washers.

10. **Examine brush faces** for chipped toes or and for heat cracks. Replace any that are damaged.

11. **Check brush spring pressure** using the spring balance method. Readjust the spring pressure in accordance with the manufacturer's instructions.

12. **Check the brush shunts** to make sure they are properly secured to the brushes and holders. In some instances, if changes have occurred during the operation of equipment since installation, it may be necessary to check the following points that would not ordinarily be disturbed.

    (a) Reset brushes at the correct angle.
    (b) Reset brushes in the neutral plane.
    (c) Properly space brushes on the commutator.
    (d) Correctly stagger the brush holders.
    (e) Properly space brush holders from the commutator.
    (f) Check to ensure that the correct grade of brush recommended by the manufacturer is being used.

- **Collector rings.** The surest means of securing satisfactory operation is to maintain the slip-ring surface in a smooth and concentric condition.

1. **Check insulation resistance** between ring and shaft to detect cracked or defective bushings and collars.

2. **Clean** using a solvent cleaner and stiff brush.

3. **Check brush holder end play and staggering** to prevent grooving the rings during operation.

4. **Check for wear of rings.** When the rings have worn eccentric with the shaft, the ring face should be machined.

5. **Check commutators.** In general, sources of unsatisfactory commutation are due to either improper assembly of current collecting parts or faulty operating condition.

6. **Check commutator concentricity** with a dial gauge, if sufficient evidence indicates that the commutator is out of round. A dial indicator reading of 0.001 inch on high-speed machines to several thousandths of a inch on low-speed machines can be considered normal.

7. **Examine commutator surface** for high bars grooving, evidence of scratches, or roughness. In light cases, the commutator may be hand-stoned, but for extreme roughness, turning of the commutator in the lathe is recommended.

8. **Check for high or pitted mica** and undercut where deemed advisable.
    After conditioning a commutator, make sure it is completely clean, with every trace of copper, carbon, or other dust removed.

- **Bearings and lubrication.** The bearings of all electrical equipment should be subjected to careful inspection at scheduled periodic intervals to assure maximum life. The frequency of inspection is best determined by a study of the particular operating conditions.

    1. **Check sleeve bearings.** In the older types, the oil should be drained, the bearing flushed, and new oil added at least every year.

    2. **Check sealed sleeve bearings.** These require very little attention, oil level is frequently the only check needed for years of service.

    3. **Re oil waste-packed bearings** every 1000 hours of operation.

    4. **Check the air gap** with a feeler gauge to ensure against a worn bearing that might permit the rotor to rub the laminations. On larger machines, keep a record of these checks. Take four measurements 90 degrees apart, one of these points being the load side, and compare with readings previously recorded to permit early detection of bearing wear.

        Bearing currents on larger machines are usually eliminated by installing insulation under the pedestals or brackets. Elimination of the circulating currents prevents pitting the bearing and shaft. From a maintenance standpoint, make sure that the pedestal insulation is not short-circuited by metal thermostat or thermometer leads, or by piping.

- **Ball and roller bearings.** External inspection at the time of greasing will determine whether the bearings are operating quietly and without undue heating.

    1. **Check the condition of the bearings** and grease. The bearing and housing parts should be thoroughly cleaned and new parts added. Where special instructions regarding the type quantity of lubricant are recommended by the manufacturer, they should be followed. In all cases, strictly adhere to standard greasing practices.

- **Thrust bearings.** Established lubrication practice for sleeve bearings applies in general to thrust bearings.

CHAPTER 2

Designing Motor Feeder, Branch Circuits, Overloads, and Motor Application*

Design and Application of Overloads

The purpose of overloads, or *heaters,* as they are sometimes called, is to protect the windings of the motor. These overloads (heaters) also provide overload protection for the branch and feeder circuits.

The proper protection of motors is important to:

- minimize damage to the motor and associated equipment
- enhance safety of personnel in the area of the motors
- maximize productivity

Overloads come in two types: (1) *thermal overloads,* which are tripped when generated heat is greater than its setting. This is accomplished with a bimetallic strip, coil, and solder pot; (2) *solid-state overloads,* which are opened when current flow through the overload device is greater than the setting of the device. This is accomplished with the doping of a semi-conductor material (see more details in the book *Industrial Electronics.* These overloads are very versatile because the trip setting is adjustable and reliable.

The slower the motor turns (compared to the design speed), the more current it will draw from the line. Devices used to protect the motor from overheating are calibrated according to motor current.

Principles of Overload Operations

- Thermal
- Bimetallic
- Relay
- Electronic

Thermal Principle   The thermal principle is the principle of a heater to heat a melting alloy (eutectic alloy) designed to become fluid at a certain temperature. Often this type of overload protection (OL) is called a *solder pot* because the alloy resembles solder, which becomes fluid when heated and rehardens when cooled.

*All sections are adapted from the National Electrical Code, NFPA 70, 1996 edition except section 6-Siemens.

As the heater gets hot because of too much motor current, it melts the alloy and allows the inner shaft and ratchet wheel to turn. This allows a set of contacts to open, which opens the control circuit, interrupting the power circuit to the motor. When the current to the motor is interrupted, the motor and the solder pot begin to cool. A manual reset is wired where you physically have to push the contacts back to the closed position. Remember, resetting this type of overload too soon may cause the shaft to spin and not reset the solder pot.

Bimetallic Expansion Principle Bimetallic expansion is also called *bimetal strip overload relay.* The heater is placed in the overload part of the starter. The heat directly affects a strip of metal—actually, two dissimilar metals adhered together. The two metals have different coefficients of expansion when heat is applied—that is, they expand at different rates when heated. This causes the strip to bend as one metal begins to expand faster than the other (similar to a home thermostat). As the metal bends, it reaches a point where it will touch off a set of contacts to open the control circuit to the motor. An advantage of this particular style of OL is that automatic reset is available. As the bimetal strip cools, it could allow the contacts to remake and allow the motor to restart.

Another style of thermal protection looks like a disk. The disk is warmed by a circular heater. The disk is a bimetallic device that changes its shape from concave to convex when heated the proper amount. The change causes a set of contacts to open, as do the other styles of thermal elements.

In the case of *overload relays,* where current causes a mechanical device to open and prevent further current flow to the motor, there is a time lag between heat sensing and OL trip. This time lag is desirable. It allows the initial starting inrush current to the motor to flow without tripping the overload device. Without this delay, every time the motor attempts to start the OL would trip. However, the delay is too long to protect the motor against short circuits and ground faults.

Electronic Overload*

The use of electronic components, whether they be discrete or integrated solid-state electronics (ASICs or microprocessors), in the construction of overload relays has resulted in enhanced protection, improved features, and communications being offered by those products. Enhanced protection includes:

- phase loss protection
- phase imbalance
- phase sequence
- jam protection
- ground fault (earth fault) protection

Improved features from electronics include:

- increased accuracy and repeatability
- lower heat generation and energy usage
- wide current adjustment range
- selectable trip class
- control functions

*Written courtesy of Rockwell Automation/Allen Bradley.

The primary cause of motor failure is excessive heat, which is caused by excess current (current greater than the normal motor full-load current), high ambient temperatures, and poor ventilation of the motor. In general, a *single* motor protective device cannot protect the motor from excessive heat due to all three of those causes. Currents greater than normal motor full-load current can be caused by high inertia loads, such as loaded conveyors, locked rotor conditions, low voltage, phase failure, and phase imbalance. If a motor is continuously overheated by only 10 degrees, its life can be reduced by as much as 50 percent.

Selectable Trip Class Solid-state electronics incorporated in the motor protective device enable selectable trip class to be incorporated into its design. Selectable trip class can be in the form of selecting class 10, 20, or 30, as with traditional overload relays. With some microprocessor-based products, however, the selectable trip class is virtually infinite, where the trip time of the motor protective device can be programmed to any specific time that is suitable for the application, whether it be 1 second, 10 seconds, 17 seconds, or 99 seconds. With traditional electromechanical overload relays, individual heater elements with specific trip classes must be purchased separately, in addition to the overload relay, to obtain selectable trip class. The primary advantage of the selectable trip class is that customers can minimize their stock and utilize a single motor protective device for standard motor starting applications, as well as for special motor starting applications where there might be a long motor starting time that would require a slower trip class. An example of this type of application would be a centrifuge, or a pump that was required to pump a very thick fluid.

Setting and Repeat Tripping Accuracy Electronic motor protective devices, on the other hand, can offer setting and repeat tripping accuracies of 2.5 percent. Setting the overload protective device is typically accomplished by using potentiometers, DIP switches, or keypad entry. Repeat accuracy of electronic motor protective devices can be as low as 1 percent, and is achieved by precise manufacturing tolerances of the various electronic components that make up motor protective devices—the resistors, capacitors, transistors, ASICS, and microprocessors.

Wide Current Adjustment Range Solid-state motor protective devices have adjustment ranges anywhere from 3.2:1 up to 9:1. With a 9:1 adjustment range, obviously they can eliminate even more overload relays and reduce the number of different components that are required for a wide range of applications.

Special Control Applications Electronic motor protective devices or motor starters that incorporate them are extremely well-suited for special motor applications—those applications that require different performance than typical direct online motor starting.

In wye-delta (star-delta) starting applications, solid-state motor protective devices can be programmed to switch from a wye (star) to a delta wiring configuration as soon as the starting current has dropped to the rated value and the motor has reached its normal speed in the wye configuration.

The programmability of some solid-state motor protective devices also provides a great deal of flexibility to vary the current levels of both the wye (star) and the delta configuration, as well as the starting time.

In summary, the motor protective device is providing the performance and functionality of separate wye-delta (star-delta) timers that would be required in the automatic operation of wye-delta (star-delta) starters.

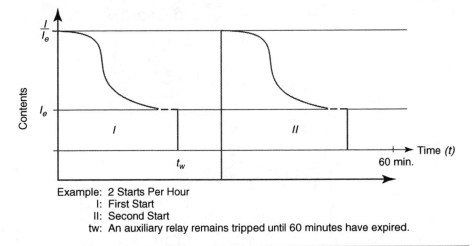

Example: 2 Starts Per Hour
        I: First Start
       II: Second Start
      tw: An auxiliary relay remains tripped until 60 minutes have expired.

Figure 2–1   Limited number of starts per hour; Courtesy of Rockwell Automation/ Allen-Bradley.

Another control function that electronic motor protection relays can provide is limiting the number of starts over a specific period of time. Figure 2–1 demonstrates time current characteristics of limiting the number of starts of a given motor.

In many critical process applications, the ability to restart a motor even after an overload condition is mandatory. Examples include mines and tunnels, where fresh air is always required, applications in which a process must not be interrupted for too long a period of time, or the product will be ruined.

The motor protective device can be configured to allow the necessary inrush current, starting time, and heat generation that would be required to get the motor started and prevent the relay from nuisance tripping. Similar to the warm start characteristics of the motor, there may also be installations where emergency starting of a motor is required, even though the overload relay has tripped due to an overcurrent condition.

Communication

A starter's or motor protective device's ability to communicate information back to a main processor or controller provides a complete spectrum of new opportunities to optimize processes and maximize productivity. While communication via a network bus to a PLC or personal computer is the most common form of communication, other methods of communication are also available. Figure 2–2 and Figure 2–3 show typical installations where multiple starters may be installed on a control network communicating to a PLC and personal computer or communicating to an interface module that would be installed in the relative vicinity of the starter.

With many solid-state motor protective devices, communication to a control module and communication with a PLC can take place simultaneously. To maximize the productivity of a particular application, the data communicated by the motor protective device can be used to monitor and manage the process to prevent conditions where the device would trip, interrupting the production. Many motor protective devices have pre-warning levels that are associated with the various causes of trip conditions. These pre-warning levels may

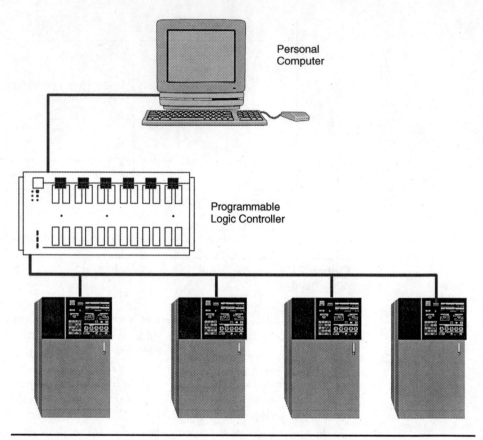

Personal
Computer

Programmable
Logic Controller

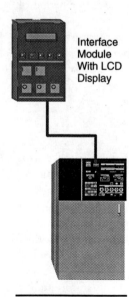

Interface
Module
With LCD
Display

Figure 2–2   Typical connections; Courtesy of Rockwell Automation/Allen-Bradley.

be for ground fault conditions, starting time conditions, phase imbalance conditions, underload conditions, and high overload or jam conditions. These pre-warning levels can be assigned to output relays that energize alarms or provide information to process operators, advising them to change the flow rate or process rate of the application to prevent the motor protective device from tripping. This process modification may include:

- slowing down a conveyor system
- reducing the flow rate in a pumping system
- unplugging or cleaning a filter
- replacing bearings or belts
- replacing cutting tools

Figure 2–3
Interface module;
Courtesy of Rockwell
Automation/
Allen-Bradley.

Finally, processes and motor utilization can be optimized with the increased functionality of solid-state motor protective devices. The ability to set pre-warnings on the motor protective device and to monitor the actual motor application parameters enables the operators to utilize the motor to its maximum capacity. This helps to ensure the greatest production output, as well as helping to ensure that the motor will not be damaged by overcurrent or overheating conditions.

Design and Application of Branch Circuits

The Principle of Overcurrent Protection

The purpose of branch circuit protection for motors is to provide ground fault and short circuit protection for motor circuit conductors and equipment. There is no need to provide extra overload protection in a branch circuit because it is provided by the starter or internal with the motor windings. Duration of the fault determines the type overcurrent protection needed in the motor circuit. Ground faults and short circuits last for only a short period of time, this explains why the overcurrent protection can be designed to be much larger than the full-load current rating of the motor. Overload faults lasts for longer periods of time than short circuits and ground faults. Overload protection must be sized closer to the full-load rating of the motor. The primary purpose of overload protection is to protect the windings of the motor. The term *overcurrent* refers to providing ground fault and short circuit protection, which *can* include—but does not always include—overload protection. The term *overload* refers to overload conditions but not short circuit or ground fault condition.

Typical Methods of Branch Circuit Overcurrent Protection

Fuse
> Nontime Delay Fuse
> Dual Element Time Delay Fuse

Circuit Breakers
> Instantaneous Trip Breaker
> Inverse Time Breaker

Sizing Branch Circuit Conductors

To properly size the branch circuit conductors feeding one motor at one speed, you must first determine the full-load current (FLC) rating of the motor. This (FLC) can be obtained from Tables 430–148 through Table 430–150 of the National Electrical Code. Refer to Table M in the appendix. These tables are based on different power factors and voltages. You may see a difference between the nameplate FLC and the FLC shown on the table. Use the table rating in place of the nameplate rating, as required by article 430–6 of the NEC for continuous loads. Short time, intermediate, and varying duty use the nameplate rating in accordance with article 430–22 exception of the NEC.

After you have determined the full-load current of the motor, multiply the FLC by 125 percent. The resultant of this multification will give you the amperage needed to size your branch circuit conductors. See *Example #1a* for an example of this calculation. Refer to *Chart #1(b) (c) (d)* and *Tables M, P, Q,* and *R* in the appendix for conductor sizes already given for the horespower and voltage given. This chart is a quick reference for those who do not want to work with the calculation.

When multispeed motors are used the conductor current on the line side of the controller must be based on the largest current shown on the nameplate. The current rating for conductors from the load side of the controller to the motor must be based on the winding current of the motor. When conductors are used for short time, intermediate, periodic or varying duty the conductor ampacity can be based on the percentages given in Article 430–33(a) of the NEC which can be found in *Chart #3* in the appendix.

In the case of DC motors, the conductor ampacity for conductors located between the controller and the motor shall be based on the following:

1. 190 percent of the FLC when the single-phase half-wave type bridge rectifier is used

2. 150 percent of the FLC when the single-phase full-wave type bridge rectifier is used

Sizing the Wound Rotor Secondary Conductors

All motors must be considered to operate at continuous duty unless authorized differently. Conductors operating at continuous duty from the secondary wound rotor motor to the controller must be sized at 125 percent of the full-load current of the motor. For other than continuous duty see *Chart #3* in the appendix for percentages allowed.

In cases where the secondary resistors are separate from the controller, the size of the conductors from the controller to the resistor shall be based on *Table N* in the appendix. These calculations do not take into consideration voltage drop. In some cases, voltage drop may need to be considered when sizing the branch circuit conductors.

Sizing Branch Circuit Overcurrent Protection

When sizing the overcurrent protection device, remember that you are primarily protecting the equipment and conductors for short circuits and ground faults. Overload protection should be provided with the controller or integral with the motor windings.

To properly size the branch circuit overcurrent protection device, you must first determine the FLC of the motor. Then you must decide what type of overcurrent device you will be using. The type of device used will be based on the type load being supplied. More detailed information is given later in this chapter. There are primarily two types of fuse protection that can be used: time delay dual element and nontime delay. There are primarily two types of circuit breakers: inverse time circuit breaker and instantaneous trip breaker. Once FLC and the OCPD have been selected, multiply the percentages given in Table 430–152 of the National Electrical Code, the resultant of the calculation will give you the calculated load. Use this calculation to size the OCPD. For standard size OCPD's available, see *Chart #2* in appendix. In cases when the calculation based on table O does not correspond to the standard size OCPD or adjustable type OCPD the next lower size can be used. However, if the lower size is not capable of carrying the load the next higher standard size is permitted.

How to Size Overcurrent Protection When Starting Current Opens Motor Circuits

When the calculation derived from the preceding given method is not adequate to start the motor the following can be substituted.

- The size of the *nontime-delay fuse* less than 600 volts can be increased by no more than 400 percent of the motor full-load current.

- The size of the *time-delay (dual element) fuse* can be increased by no greater than 225 percent of the motor full-load current.

- The size of the *inverse time circuit breaker* increased by no more than 400 percent of the motor full-load current for motors with a FLC of 100 amperes and no more than 300 percent of the FLC for motors with FLC of greater than 100 amperes.

- The size of *600–6000 ampere fuses* can be increased by no more than 300 percent of the motor full-load current.

- The size of the *nontime-delay fuse* less than 600 volts can be increased by 400 percent of the motor full-load current.

- The size of the *instantaneous circuit breaker* that is adjustable and part of a tested (listed) combination controller and circuit breaker can be increased by 1300 percent of the motor full-load current.

- Special *short circuit protectors* may be used as a substitute for any of the preceding OCPD's given.

- Overcurrent protection for *torque motors* shall be based on the motor nameplate. When the motor nameplate full-load current of a torque motor does not correspond with the standard sizes given in *Chart #2* of the appendix, the next higher size can be used for motors rated no greater than 800 amperes. For torque motors with a FLC greater than 800 amperes do not use the next higher standard size, you must use the equal or lower size.

For examples of overcurrent protection calculations see *Example #1b, step 4* in the appendix.

Design and Application of Feeder Circuits

The Principle of Feeder Overcurrent Protection

The purpose of feeder protection is to provide ground fault and short circuit protection for motor circuit conductors and equipment. There is no need to provide extra overload protection in a feeder circuit because this type of protection is provided by the starter or is internal with the motor windings. Duration of the fault determines the type of overcurrent protection needed in the motor circuit. Ground faults and short circuits last for only a short period of time. This explains why the overcurrent protection can be designed to be much larger than the full-load current rating of the motor.

Overload faults last for longer periods of time than short circuits and ground faults. Overload protection must be sized closer to the full-load rating of the motor. The primary purpose of overload protection is to protect the windings of the motor. The term *overcurrent* refers to providing ground fault and short-circuit protection, which *can* include (not always) overload protection. The term *overload* refers to overload conditions but not short circuit or ground fault condition.

Typical Methods of Feeder Overcurrent Protection

Fuse
    Nontime Delay Fuse
    Dual Element Time Delay Fuse

Circuit Breakers
    Instantaneous Trip Breaker
    Inverse Time Breaker

Sizing Feeder Circuit Conductors

Feeder conductors are defined as the conductors that extend from the source to the branch circuit conductors. See *Example 1b, step 2* in the appendix. To properly size the feeder conductors, add the full load currents of all motors in the circuit plus 25 percent of the full-load current rating of the largest motor in the group. In cases where two or more motors are the largest motor in the group with the same FLC, use only 25 percent of one of the motors to calculate the load. When other loads, such as lighting fixtures are part of the feeder circuit, add the other loads at 100 percent of the rating.

When one or more of the motors in the group has a *varying, short-time, intermediate,* or *periodic duty* use *Chart #3* in the appendix to determine the ampacity of other than continuous duty use motors. Once ampacity of the noncontinuous duty motors has been determined add 25 percent of the largest motor of the group. The largest motor of the group is considered to be the continuous-duty motor at 125 percent or noncontinuous motor based on *Chart #3,* whichever has the largest full-load current. In cases when the circuitry of the motors is so *interlocked* as to prevent all motors or more than one motor from being energized at the same time, the size feeder conductors shall be based on 100 percent of all motors in the group. Once the calculation has been established with the preceding procedures, see article 310–16 of the NEC or charts for ampacity of conductor sizes.

Sizing Feeder Circuit Overcurrent Protection

Feeder circuit overcurrent protection can be determined by using the following procedure.

1. Add the largest overcurrent protection device plus the full load currents of all other motors in the group. *Note:* When two motors in the group have the same rating and are also the largest motor in the group, count one as the largest overcurrent device in the group.
2. Add the full load current of the other largest motor in the group. The same applies if all motors of the group are the same size. When instantaneous trip circuit breakers are used as the branch circuits, it is assumed that the percentages given in *Table O* in the appendix have been applied before the feeder calculation has been made.
3. Size the OCPD based on the standard ratings listed on *Chart #2* and refer to *Example #1b, step 5* in the appendix.

Design and Application of Disconnecting Means

The purpose of the disconnect is to remove the motor and controller (if used) from the power source. The disconnect also must be capable of withstanding full load and locked rotor currents without causing damage to the equipment.

Sizing the Disconnect

For disconnects rated *600 or less,* the size shall be determined by multiplying the full load current of the motor by 115 percent. For disconnecting means rated *over 600 volts,* the size of the disconnecting means shall be not less than the full load current rating of the motor being energized. When *torque motors* are being used, the size of the disconnecting means can be determined by multiplying the full load current by 115 percent. When *one disconnecting means serves as the only disconnect to more than one motor* and/or other loads (lighting, heaters, etc.), the size of the can be determined by the following: Calculate all currents on the circuit (include full load and locked rotor currents). The full loads can be obtain by using the appendix *Tables M, P, Q, and R.* The locked rotor currents can be obtained by using *Tables S1* and *S2*. A disconnecting means must be capable of opening the locked rotor current at the disconnect before damage occurs to electrical equipment in the circuit during a locked rotor condition. Locked rotor ratings can be compared to the horsepower rating of the disconnecting means. The ampere rating of the disconnecting means can be derived by multiplying 115 percent by all the full load currents in the group. For motors smaller than 1/4 HP direct current motors, 1/6 HP single phase motors, 1/2 HP two phase motors, and 1/2 HP three-phase motors consider the locked rotor current to be six times the full load currents. It is important to understand the characteristics of locked rotor and full load current. The primary purpose for full load current protection is to prevent the burning of the equipment. An improper-sized disconnecting means in relation to full load current will usually burn for a long period of time before a fault is noticed. When a disconnect is improperly sized for locked rotor current, the resultant is usually an explosion and immediate damage to equipment or those close to the fault. Locked rotor current generally is a very high current (several times higher than the full load current) and last only a short period of time before the OCPD opens or damage occurs to the equipment.

Location of the Disconnect

The disconnect must deenergize the controller or motor. It must also be located in sight from the controller, machine, or motor intended to be energized or deenergized. The term *in sight from* means the disconnect must be located not more than 50 feet from the motor or controller being utilized for power control of the equipment.

For systems 600 volts or more/less (article 430–102) the controller or disconnect is not required to be in sight if it can be locked out and is properly identified with a warning label stating the location of the disconnect or controller. Also, one disconnect can be used as the sole disconnecting means of several motors if they are all next to each other. When utilized as the sole disconnecting means, this device must be capable of being locked in an open position. See Figure 2–4.

A disconnecting means must be readily accessible—cleared of quick energizing, deenergizing, or maintenance, has to climb over, move obstacles, or use ladders or chairs to gain access to disconnecting means. The disconnect cannot be located in a locked room (see Figure 2–5).

Type Disconnecting Means

Disconnecting means come in many types. All disconnects must be listed for the application being used. *Listed* disconnects have been tested and approved by a testing agency certified as a National Recognized Testing Lab by the U.S. Department of Labor. Typical types are:

1. listed *switch*—rated to match the horsepower of the motor being used
2. listed *circuit breaker*—rated in amperage equal to the motor being used
3. listed *molded case circuit switch*—rated in amperage equal to the motor being used; a nonautomatic circuit-interrupting device without overcurrent protection

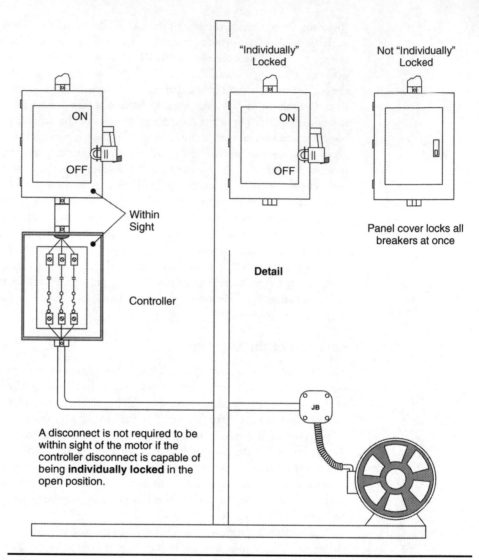

Figure 2–4 Arrangement of motor disconnecting means; Courtesy of Mike Holt, Delmar Publishers.

Stationary Motors For motors 1/8 HP and smaller, the circuit breaker can be used as the disconnecting means. For motors 2 HP and smaller rated at 300 volts and smaller, a general-use snap switch can be used in lieu of these listed types provided that the switch is rated at no less than twice the full load current of the motor (see Figure 2–6).

For DC motors greater than 40 HP or AC motors rated at greater than 100 HP a general-use type or isolating switch can be used when it is marked "DO NOT OPEN UNDER LOAD." See Figure 2–7.

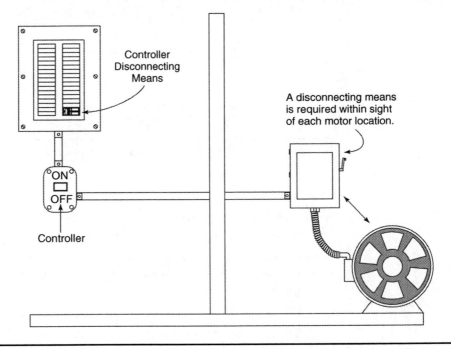

Figure 2–5 Location of motor disconnecting means; Courtesy of Mike Holt, Delmar Publishers.

Figure 2–6 Article 430–109 Exception #1½-HP or less; Courtesy of Carpenter, Delmar Publishers.

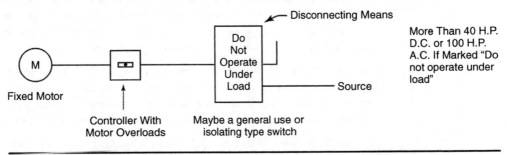

Figure 2–7 Type disconnecting means; Courtesy of Carpenter, Delmar Publishers.

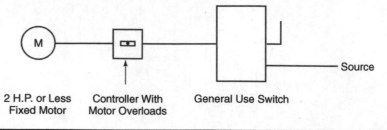

Figure 2–8   Article 430–109 Exception #2; Courtesy of Carpenter, Delmar Publishers.

On AC circuits, an *AC general-use snap switch* can be used in place of the three listed types, provided the motor is rated at 2 HP or smaller, 300 volts or smaller, and the switch has an ampere rating of 125 percent of the load current of the motor (see Figure 2–8).

For All Motors Rated from 2–100 Horsepower   When the motor is used with an auto-transformer-type controller, a general-use type switch can be used in lieu of the three listed types, given these provisions:

1. The motor is used as an exciter for generators with overload protection.
2. The controller can withstand the locked rotor current of the motor.
3. The controller provides no voltage release.
4. The overload protection must not exceed the 125 percent of the motor full load current.
5. Branch circuit disconnect fuses or inverse time circuit breaker cannot be rated at more than 150 percent of the motor full load current.

Cord-and-Plug Connected Motors   When cord-and-plug connected motors are used, the cord and plug can be used as a disconnecting means. The cord and plug must be rated no less than the motor being served.

Torque Motors   When torque motors are utilized, the disconnect can be a general snap switch in lieu of the three listed types.

Combination Motor Controller   When the combination of the motor controller and disconnect are located in the same enclosure, the instantaneous trip circuit breaker can be used as a disconnecting means. The combination controller and disconnect, just described, must be listed by a nationally recognized testing laboratory.

Switch or Circuit Breaker Used as Both Controller and Disconnecting Means   A switch or circuit breaker can be used as both the controller and the disconnect, provided the proper overcurrent protection, controller horsepower rating, and voltage rating are provided. Examples include air-break switches, inverse time breakers, and oil switches.

Design E Motors   When a design E motor is installed, a motor circuit switch can be used if the following requirements are met: (1) the switch is rated for use with a design E motor; (2) for motors rated 3–100 HP the switch must be sized 1.4 times the rating of the motor; (3) for motors rated 100 HP or more the switch must be sized 1.3 times the rating of the motor.

Enclosure Types for Disconnecting Means

Enclosures are designed to protect the electrical and mechanical equipment from different types of environment. National Electrical Manufacturers Association, (NEMA) has a design standard for different applications. A partial list and general information are given in Figure 1–43. International European Commission, (IEC) also has ratings of enclosures (see Figure 1–44 for cross reference).

Design and Application of Controller

Methods of Controller Use (see Figure 2–11a-e)

1. **Start and stop method.** The controller must control the starting and stopping of the motor by removing the rotor circuit current or electrical power delivered to the motor. See Figure 2–9*a*.

   (a) Stationary motors 1/8 HP or less with overload protection can use the branch circuit as the controller. See Figure 2–9*b* and *d*.

   (b) Portable motors of less than 1/3 HP can use the cord and plug as controller. See Figure 2–9*c*.

2. **Autotransformer.** For autotransformer schematic and designs, see Chapter 2 of this handbook. Autotransfromer controllers shall be designed to prevent the overloads from becoming inoperative during the sequence of operation.

3. **Rheostats.** Rheostats must not allow the contact arm to rest on intermediate segments. The point or place on which the arm rests must not have any electrical connection with the resistor while in the start position.

   DC motors operating from a constant voltage must have automatic devices that will open the power supply before the motor falls to less than one-third its normal rate.

4. **Horsepower rating at the application voltage.** The controller must have a horsepower rating greater than the horsepower rating of the motor.

A stationary motor less than 2 HP and 300 volts may use a general-use switch with an ampere rating equal to or greater than twice the full-load current rating of the motor.

Branch-circuit inverse-time circuit breakers can be used as a controller when matched with the proper ampere rating (see Figure 2–11*d*). Other than general-use switches may be used with a rating 80 percent of the full-load current of the motor (see Figure 2–9*e*).

Motor controllers for torque motors must have a continuous-duty, full-load current rating equal to or greater than nameplate current rating of the motor. In cases where the motor controller is rated in horsepower but not marked with a current rating, the equivalent current rating can be determined from the horsepower rating by using appendix *Tables M, P, Q, R, S1,* and *S2* or 430–147, 430–148, 430–149, or 430–150, 1996 edition of the National Electrical Code, NFPA–70.

Voltage Rating   A controller voltage rating must not exceed the applied voltage of any conductor feeding the controller. Where the controller serves also as a disconnecting means, it must open all ungrounded conductors to the motor.

Grounded Conductors   When one pole of the controller has a permanently grounded conductor (i.e., corner grounded delta), provided the controller is so designed that the pole in

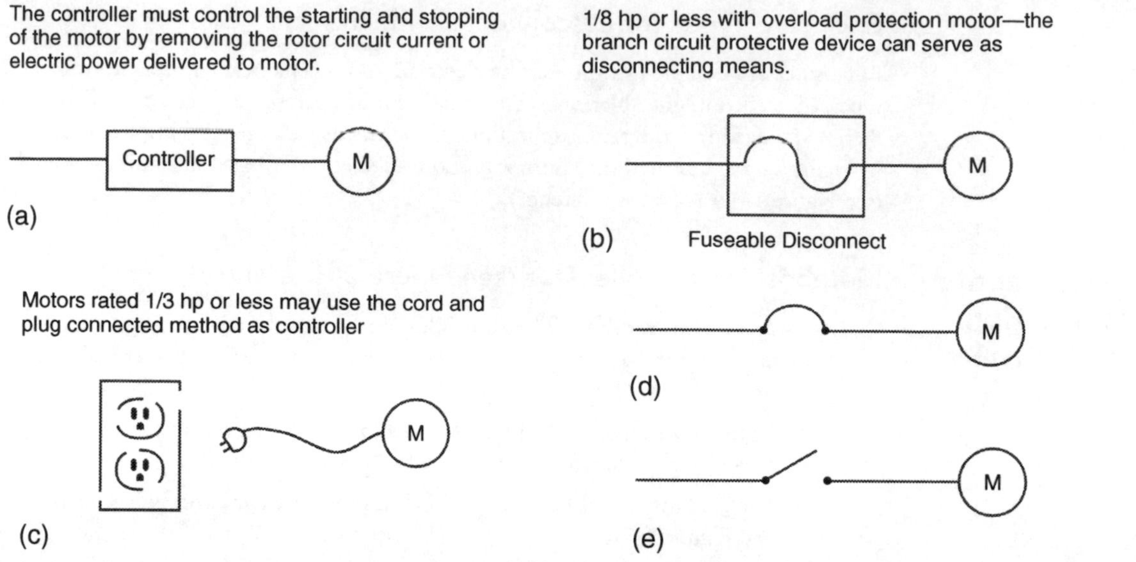

The controller must control the starting and stopping of the motor by removing the rotor circuit current or electric power delivered to motor.

(a)

1/8 hp or less with overload protection motor—the branch circuit protective device can serve as disconnecting means.

(b) Fuseable Disconnect

Motors rated 1/3 hp or less may use the cord and plug connected method as controller

(c)

(d)

(e)

Figure 2–9a-e Examples of controller installations.

the grounded conductor cannot be opened without simultaneously opening all conductors of the circuit.

Number of Motors Served by Each Controller   Each motor should be provided with an individual controller. In cases when the motor is rated 600 volts or less, a single controller can be used to control a group of motors provided the controller is rated not less than the sum of the horsepower ratings of all of the motors of the group. The single controller must be located in one room or must control one machine and have one overcurrent protection device serving as the only protection.

Speed Limitation

Speed limitation is required for the following types of motors.

1. separately excited DC motors
2. series motors
3. motor-generators and converters

When the inherent characteristics of the motor provides speed limitation external speed control is not required. Also, if a qualified operator is present to prevent the overspeed situation, manually external speed control is not required.

Combination Fuseholder and Switch as Controller

The rating of a fusable disconnect switch must correspond to the motor overload protection (see Figure 2–10).

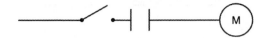

Figure 2–10 Combination disconnect and overload.

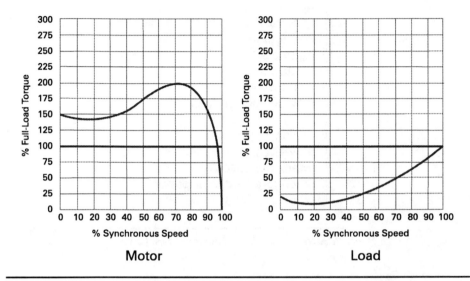

Figure 2–11 Matching AC motors to the load; Courtesy Siemens Energy and Automation, Inc.

How to Properly Apply a Motor to the Intended Load

Matching AC Motors to the Load

One way to evaluate whether the torque capabilities of a motor meet the torque requirements of the load is to compare the motor's speed-torque curve with the speed-torque requirements of the load. See Figure 2–11.

Load Characteristics Tables

To find the torque characteristics, use a table similar to the partial one shown in Figure 2–12. NEMA publication MG 1 is one source of typical torque characteristics.

Calculating Load Torque

The most accurate way to obtain torque characteristics of a given load is to obtain them from the equipment manufacturer. A simple experiment can be set up to show how torque of a given load can be calculated. In the following illustration, a pulley is fastened to the shaft of a load that a motor is to drive. A cord is wrapped around the pulley with one end connected to a spring scale. The torque can be calculated by pulling on the scale until the shaft turns and then noting the reading on the scale. The force required to turn the shaft, indicated by the scale, times the radius of the pulley equals the torque value. Remember that the radius is measured from the center of the shaft to the center of the pulley. If the radius were 1 foot, for example, and the force required to turn the shaft were 10 pounds, the torque requirement is 10 ft lb. The amount of torque required to turn the connected load can vary at different speeds (see Figure 2–13).

Load Description	Load Torque as % Full-Load Drive Torque		
	Break-away	Accel-erating	Peak Running
Actuators:			
Screw-down (rolling mills)	200	150	125
Positioning	150	110	100
Agitators			
Liquid	100	100	100
Slurry	150	100	100
Blowers, centrifugal:			
Valve closed	30	50	40
Valve open	40	110	100
Blowers, positive displacement, rotary, bypassed	40	40	100
Calenders, textile or paper	75	110	100

Figure 2–12   Load characteristics; Courtesy Siemens Energy and Automation, Inc.

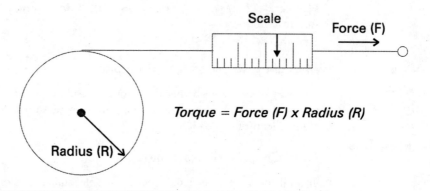

Figure 2–13   Calculating load torque; Courtesy Siemens Energy and Automation, Inc.

Centrifugal Pump   When a motor accelerates a load from zero to full-load speed the amount of torque it can produce changes. At any point during acceleration and while the motor is operating at full-load speed, the amount of torque produced by the motor must always exceed the torque required by the load. In the following example a centrifugal pump has a full-load torque of 600 ft lb. This is equivalent to 200 HP. The centrifugal pump only requires approximately 20 percent of full-load torque to start. The torque dips slightly after it is started, and then increases to full-load torque as the pump comes up to speed. This is typically defined as a variable torque load (see Figure 2–14).

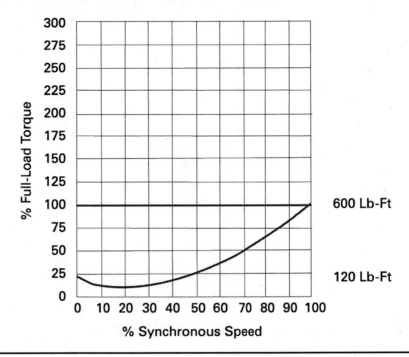

Figure 2–14   Centrifugal pump; Courtesy Siemens Energy and Automation, Inc.

A motor has to be selected that can start and accelerate the centrifugal pump. By comparing a 200 HP NEMA B motor curve to the load curve, it can be seen that the motor will easily start and accelerate the load (see Figure 2–15).

Screw-Down Actuator

In the following example, a screw-down actuator is used. The starting torque of a screw down actuator is approximately 200 percent of full-load torque. Comparing the load's requirement with the NEMA design B motor of equivalent horsepower, it can be seen that the load's starting torque requirement is greater than the motor's capability. The motor will not start and accelerate the load (see Figure 2–16).

One solution would be to use a higher horsepower NEMA B motor. A less expensive solution might be to use a NEMA D motor of the same horsepower requirements as the load. A NEMA D motor would easily start and accelerate the load (see Figure 2–17).

The motor selected to drive the load must have sufficient torque to start, accelerate, and run the load. If, at any point, the motor cannot produce the required torque, the motor will stall or run in an overloaded condition. This will cause the motor to generate excess heat and (usually) exceed current limits, causing protective devices to remove the motor from the power source. If the overload condition is not corrected, or the proper motor is not installed, the existing motor will eventually fail.

NEMA Design   Standard Motor Designs: NEMA Motor Characteristics

Motors are designed with certain speed-torque characteristics to match speed-torque requirements of various loads.

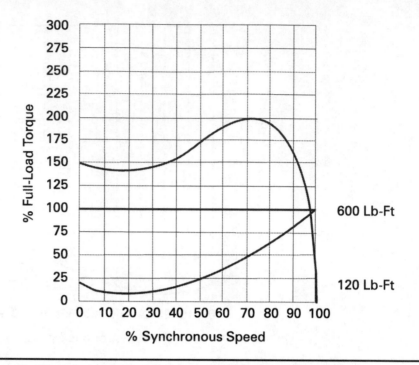

Figure 2–15   Motor selection Courtesy Siemens Energy and Automation, Inc.

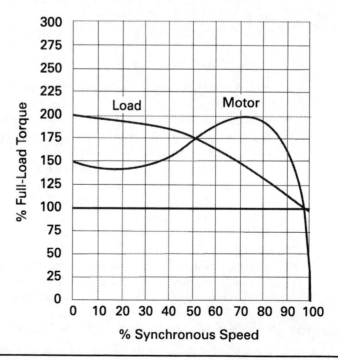

Figure 2–16   Screw-down actuator; Courtesy Siemens Energy and Automation, Inc.

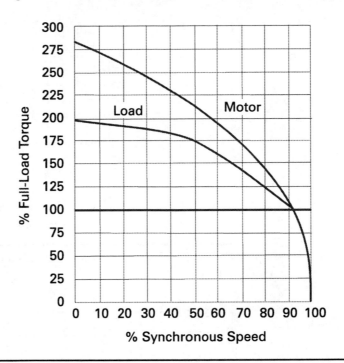

Figure 2–17 Using higher rated horsepower motor; Courtesy Siemens Energy and Automation, Inc.

The four standard NEMA designs are NEMA A, NEMA B, NEMA C, and NEMA D. NEMA A is not used very often. NEMA B is most commonly used. NEMA C and NEMA D used for specialized applications. A motor must be able to develop enough torque to start, accelerate, and operate a load at rated speed. Using a sample 30 HP, 1765 RPM motor, torque (I) can be calculated by transposing the formula for horsepower.

$$HP = T \times RPM \div 5250$$

$$T = HP \times 5250 \div RPM$$

$$T = 30 \times 5250 \div 1765$$

$$T = 89.2 \text{ ft lb}$$

Figure 2–18 shows the relationship between speed and torque the motor produces from the moment of start until the motor reaches full-load torque at rated speed. This graph represents a NEMA B motor.

To understand starting torque, see Figure 2–19. Starting torque (point *A* on the graph) is also referred to as locked rotor torque. This torque is developed when the rotor is held at rest with rated voltage and frequency applied. This condition occurs each time a motor is started. When rated voltage and frequency are applied to the stator, there is a brief amount of time before the rotor turns. At this instant a NEMA B motor develops approximately 150 percent of its full-load torque. A 30 HP, 1765 RPM motor, for example, will develop approximately 133.8 ft lb of torque.

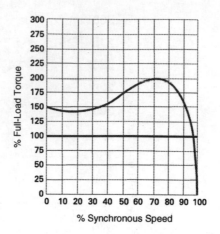

Figure 2–18 Speed-torque curve for NEMA B motor; Courtesy Siemens Energy and Automation, Inc.

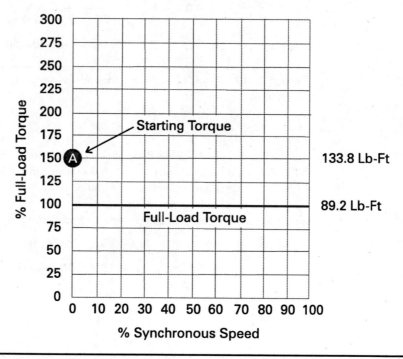

Figure 2–19 Starting torque; Courtesy Siemens Energy and Automation, Inc.

For *accelerating and breakdown torque* see Figure 2–20. The magnetic attraction of the rotating magnetic field will cause the rotor to accelerate. As the motor picks up speed torque decreases slightly until it reaches point *B* on the graph. As speed continues to increase from point *B* to point *C* torque increases until it reaches its maximum at approximately 200 percent. This torque is referred to as accelerating or pull-up torque. Point *C* is the maximum torque a motor can produce. At this point a 30 HP motor will develop

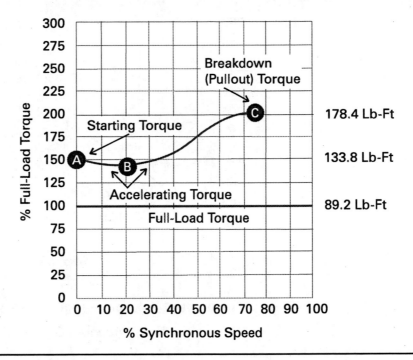

Figure 2–20   Accelerating and breakdown torque; Courtesy Siemens Energy and Automation, Inc.

approximately 178.4 ft lb of torque. If the motor were overloaded beyond the motor's torque capability, it would stall or abruptly slow down at this point. This is referred to as breakdown or pullout torque.

Full-load torque decreases rapidly as speed increases beyond breakdown torque (point C), until it reaches full-load torque at a speed slightly less than 100 percent synchronous speed. Full-load torque is the torque developed when the motor is operating with rated voltage, frequency, and load. The speed at which full-load torque is produced is the slip speed or rated speed of the motor. Recall that slip is required to produce torque. If the synchronous speed of the motor is 1800 RPM and the amount of slip is 1.9 percent, the full-load rated speed of the motor is 1765 RPM. The full-load torque of the 30-HP, 1765-RPM motor is 89.2 ft lb. NEMA B motors are general-purpose, single-speed motors suited for applications that require normal starting and running torque, such as conveyors, fans, centrifugal pumps, and machine tools (see Figure 2–21).

Starting—or locked rotor—current and full-load current are measured from the supply line at rated voltage and frequency with the rotor at rest. Full-load current is the current measured from the supply line at rated voltage, frequency, and load with the rotor up to speed. Starting current is typically 600 to 650 percent of full-load current on a NEMA B motor. Starting current decreases to rated full-load current as the rotor comes up to speed (see Figure 2–22).

NEMA A motor sets limits of starting (locked rotor) current for NEMA design B motors. When special load torque or load inertia requirements result in special electrical

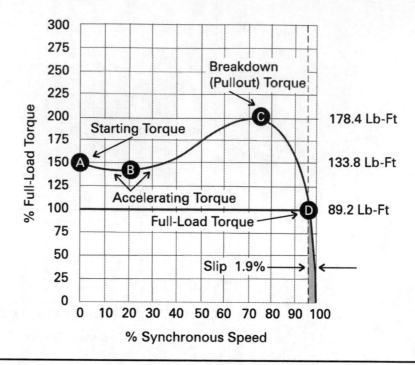

Figure 2–21   Full-load torque; Courtesy Siemens Energy and Automation, Inc.

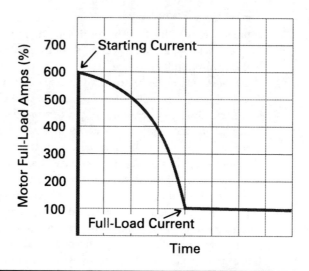

Figure 2–22   Starting current and full-load current; Courtesy Siemens Energy and Automation, Inc.

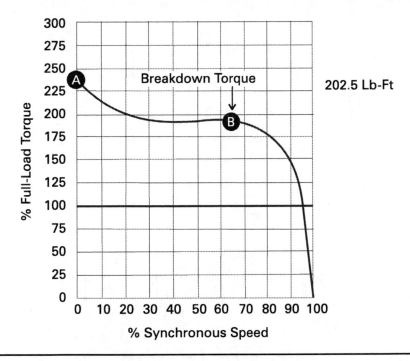

Figure 2–23 Starting characteristics of a NEMA C motor; Courtesy Siemens Energy and Automation, Inc.

designs that will yield higher locked rotor current (LRA), NEMA design a may result. This designation also cautions the selection of motor control components to avoid tripping protective devices during longer acceleration times or higher than normal starting current.

NEMA C motor starting torque is approximately 225 percent. A NEMA C, 1765 RPM, 30 HP motor will develop approximately 202.5 ft lb of starting torque. Hard-to-start—applications such as plunger pumps, heavily loaded conveyors, and compressors—require this higher starting torque. Slip and full-load torque are about the same as a NEMA B motor. NEMA C applies to single-speed motors from approximately 5 HP to 200 HP (see Figure 2–23).

NEMA D motor starting torque is approximately 280 percent of the motor's full-load torque. A NEMA D, with a full-load rated speed of 1765 RPM, and 30 HP will develop approximately 252 ft lb of starting torque. Very-hard-to start applications—such as punch presses, cranes, hoists, and oil well pumps—require this high starting torque. NEMA D motors have no true breakdown torque. After initial starting torque is reached, torque decreases until full-load torque is reached. NEMA D motors typically are designed with either 5 to 8 percent slip or 8 to 13 percent slip (see Figure 2–24).

Multispeed and ASD (adjustable speed drive) are specialized motors uniquely designed for specific load requirements. NEMA design classifications are not applicable to these specialized motors.

Various special configurations of motor controls are selected when starting or accelerating torques must be more accurately controlled, or when starting current must be limited. In the cases of part winding start or wye-delta start, the motor windings must be designed

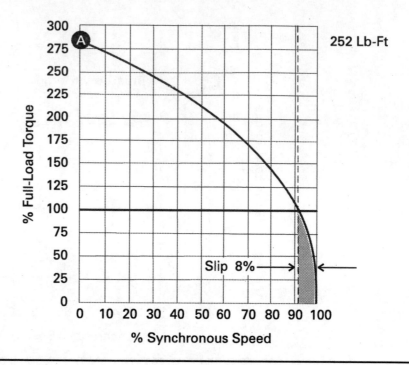

Figure 2–24 Starting torque of NEMA D motor; Courtesy Siemens Energy and Automation, Inc.

with unique connections for the special controls. In cases such as reduced voltage auto-transformer or electronic soft starts, relatively standard motors may be approved for these special applications.

Matching Motor to Load

Starting torque, breakdown torque, and full-load torque are based on speed-torque curves obtained from the driven equipment manufacturer. There are, however, some minimum torques that all *large AC motors* must be able to develop. These are specified by NEMA.

- locked rotor torque ≥ 60% of full-load torque
- pull-up torque ≥ 60% of full-load torque
- maximum torque ≥ 175% of full-load torque

Above NEMA motors require the same adjustment for altitude and ambient tempera-ture as integral frame size motors. When a motor is operated above 3,300 feet, a higher-class insulation should be used or the motor should be derated. Above NEMA motors with class B insulation can easily be modified for operation in an ambient temperature between 40° C and 50° C. Above 50° C requires special modification at the factory.

Matching AC Motors and Drives

Many applications require variable speed of an AC motor. The easiest way to vary the speed of an AC induction motor is to use an AC drive to vary the applied frequency. Operating a motor at other than the rated frequency and voltage affects motor current and torque.

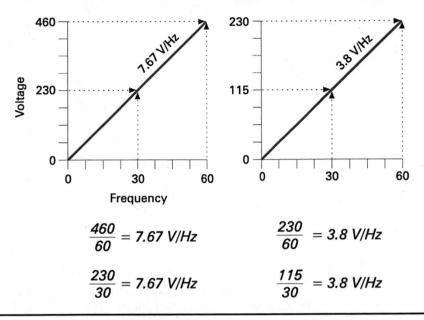

$$\frac{460}{60} = 7.67 \ V/Hz \qquad \frac{230}{60} = 3.8 \ V/Hz$$

$$\frac{230}{30} = 7.67 \ V/Hz \qquad \frac{115}{30} = 3.8 \ V/Hz$$

Figure 2–25   Constant torque; Courtesy Siemens Energy and Automation, Inc.

A ratio exists between voltage and frequency. This ratio is referred to as volts per hertz (V/Hz). A typical AC motor manufactured for use in the United States is rated for 460 VAC and 60 Hz. The ratio is 7.67 volts per hertz. Not every motor has a 7.67 V/Hz ratio. A 230-volt, 60-Hz motor, for example, has a 3.8 V/Hz ratio.

$$460 \div 60 = 7.67 \ V/Hz \qquad 230 \div 60 = 3.8 \ V/Hz$$

Flux (Φ), magnetizing current (I^M), and torque are all dependent on this ratio. Increasing frequency (F) without increasing voltage (E), for example, will cause a corresponding increase in speed. Flux, however, will decrease causing motor torque to decrease. It can be seen that torque ($T = kw$) is directly affected by flux (Φ). Torque is also affected by the current resulting from the applied load, represented here by kw. Magnetizing current (I^M) will also decrease. A decrease in magnetizing current will cause a corresponding decrease in stator or line (I^S) current. These decreases are all related and greatly affect the motor's ability to handle a given load.

AC motors running on an AC line operate with a constant flux (Φ) because voltage and frequency are constant. Motors operated with constant flux are said to have *constant torque*. Actual torque produced, however, is determined by the demand of the load.

$$T = k\Phi \div w$$

An AC drive is capable of operating a motor with constant flux (Φ) from approximately zero (0) to the motor's rated nameplate frequency (typically 60 Hz). This is the constant torque range. As long as a constant volts-per-hertz ratio is maintained, the motor will have constant torque characteristics. AC drives change frequency to vary the speed of a motor and voltage proportionately to maintain constant flux. The graphs in Figure 2–25 illustrate the volts-per-hertz ratio of a 460-volt, 60-Hz motor and a 230-volt, 60-Hz motor. To operate the 460-volt motor at 50 percent speed with the correct ratio, the applied voltage and frequency would be 230 volts, 30 Hz.

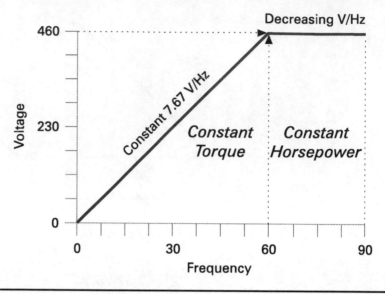

Figure 2–26   Constant horsepower; Courtesy Siemens Energy and Automation, Inc.

To operate the 230-volt motor at 50 percent speed with the correct ratio, the applied voltage and frequency would be 115 volts, 30 Hz. The voltage and frequency ratio can be maintained for any speed up to 60 Hz. This usually defines the upper limits of the constant torque range.

Constant Horsepower

Some applications require the motor to be operated above base speed. The nature of these applications requires less torque at higher speeds. Voltage, however, cannot be higher than the rated nameplate voltage. This can be illustrated using a 460-volt, 60-Hz motor. Voltage will remain at 460 volts for any speed above 60 Hz. A motor operated above its rated frequency is operating in a region known as *a constant horsepower*. Constant volts per hertz and torque is maintained up to 60 Hz. Above 60 Hz the volts-per-hertz ratio decreases, with a corresponding decrease in torque (see Figure 2–26).

Frequency	V/Hz
30 Hz	7.67
60 Hz	7.67
70 Hz	6.6
90 Hz	5.1

Flux Φ and torque T decrease:

$$\Phi = E \div F$$

$$T = k\,\Phi \div w$$

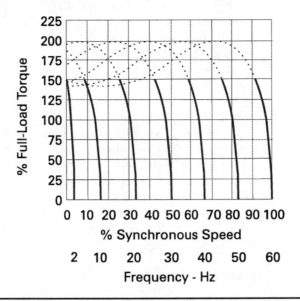

Figure 2–27 Reduced voltage and frequency starting; Courtesy Siemens Energy and Automation, Inc.

Horsepower remains constant as speed (*N*) increases and torque decreases in proportion. The following formula applies to speed in revolutions per minute (RPM).

$$\text{HP (remains constant)} = T \text{ (decreases)} \times N \text{ (increases)} \div 5250$$

Reduced Voltage and Frequency Starting

A NEMA B motor that is started by connecting it to the power supply at full voltage and full frequency will develop approximately 150 percent starting torque and 600 percent starting current. AC drives start at reduced voltage and frequency. The motor will start with approximately 150 percent torque and 150 percent current at reduced frequency and voltage. The torque-speed curve shifts to the right as frequency and voltage are increased. The dotted lines on the torque-speed curve illustrated Figure 2–27 represent the portion of the curve not used by the drive. The drive starts and accelerates the motor smoothly as frequency and voltage are gradually increased to the desired speed. An AC drive, properly sized to a motor, is capable of delivering 150 percent torque at any speed, up to speed corresponding to the incoming line voltage. The only limitations on starting torque are peak drive current and peak motor torque, whichever is less.

Some applications require higher than 150 percent starting torque. A conveyor, for example, may require 200 percent rated torque for starting. If a motor is capable of 200 percent torque at 200 percent current, and the drive is capable of 200 percent current, then 200 percent motor torque is possible. Typically drives are capable of producing 150 percent of drive nameplate rated current for one (1) minute. If the load requires more starting torque than a drive can deliver, a drive with a higher current rating would be required. It is appropriate to supply a drive with a higher continuous horsepower rating than the motor when high peak torque is required.

AC drives often have more capability than the motor. Drives can run at higher frequencies than may be suitable for an application. Above 60 Hz the V/Hz ratio decreases and the motor cannot develop 100 percent torque. In addition, drives can run at low speeds; however, self-cooled motors may not develop enough air flow for cooling at reduced speeds and full load. Each motor must be evaluated according to its own capability before selecting it for use on an AC drive.

Harmonics, voltage spikes, and voltage rise times of AC drives are not identical. Some AC drives have more sophisticated filters and other components designed to minimize undesirable heating and insulation damage to the motor. This must be considered when selecting an AC drive/motor combination. Motor manufacturers will generally classify certain recommended motor selections based on experience, required speed range, type of load torque, and temperature limits.

A high-efficiency motor with a 1.15 service factor is recommended when used on an AC drive. Due to heat associated with harmonics of an AC drive, the 1.15 service factor is reduced to 1.0.

Location of Motors

Motors must be accessible for maintenance and ventilation purposes. If proper ventilation is not maintained, the motor's life will be shorter and a potential fire hazard exist. Proper maintenance can also extend the life of the motor. However, if the motor is not accessible, the maintenance cannot be performed. It is important to notice the location of maintenance points (lubrication fittings, bearings, brushes, etc.) when installing the motor.

For motors installed in hazardous locations see Chapter 1 for installation and maintenance requirements.

Motors Over 600 Volts

High-voltage motor circuits must be designed to prevent ground faults, short circuits, and overloads to equipment and conductors associated with the motor operation. In some cases the motor must run—even under fault conditions—to prevent a greater hazard if the motor would shut down. When no ground fault, short circuit, or overload is provided, because it would cause a hazard, the motor circuit must be interlocked with an alarm or enunciator to notify those supervising the operation that a fault has occurred.

Overload Protection

All high-voltage motors must have overload protection. Thermal protection, integral with the motor or external sensing device or both, can be used as overload protection. The overloads must not be the automatic reset type if restarting could cause a hazard.

Ground Fault and Short-Circuit Protection Devices (OCPD)

Circuit breakers or fuses can serve as overload, ground fault, short-circuit protection and disconnecting means provided it opens all ungrounded (phase) conductors. The OCPD device used must not be the automatic reset type if restarting could cause a hazard.

Sizing (Ground Fault, Short Circuit) Overcurrent Protection

All disconnecting means, branch circuit, and overload devices must be sized no smaller than 115 percent of the trip current protection of the overload relays.

CHAPTER 3

Principle, Design, and Troubleshooting Single-Phase Motors*

Single-Phase Motors

In polyphase motors, the correct mechanical placement of the multiple windings in the stator, together with the phase relationships between the supply voltages, produce a uniformly rotating stator magnetic field. With a single source (single phase) of AC voltage connected to a single winding, a stationary flux field is created that pulsates in strength as the AC voltage varies but does not rotate. Consequently, if a stationary rotor is placed in this stationary stator field, it will not rotate. If the rotor is spun by hand, artificially creating relative motion between the rotor winding and the stator field, it will pick up speed and run. A single-phase, single-winding motor will run in either direction if started by hand, but it will not develop any starting torque.

What is needed is a second (run) winding, with currents out of phase with the original or main (start) winding, to produce a net rotating magnetic field, as is the case with the three-phase induction motor. These start windings, together with other components such as capacitors, relays, and centrifugal switches, make up the starting circuit (see Figure 3–1). They have varying effects on motor starting and running performance.

Split-Phase Motors

Design and Operating Principles of Split-Phase Motors

Understanding the split-phase motor operation and design is essential to understanding the operation of single-phase motors.

Because the motor has only a single AC sine wave applied to it, an artificial means of creating two voltages out of phase with each other is needed. The motor is designed to split a single sine wave voltage into two waveforms and to use the displaced voltage to create a rotating magnetic field in the stator. To do this, the stator is wound with coils of two different sizes of wire and different number of turns or loops in the stator coils.

One coil (the running, or main winding) is a physically larger diameter wire than the other coil (the starting, or auxiliary winding). The running winding is placed into the stator first so that it is deeper in the slots. This has two effects on the coil compared to the starting winding. The running winding has less resistance because it is a large diameter

*Adopted with permission from Alerich, *Electricity 4,* Delmar Publishers.

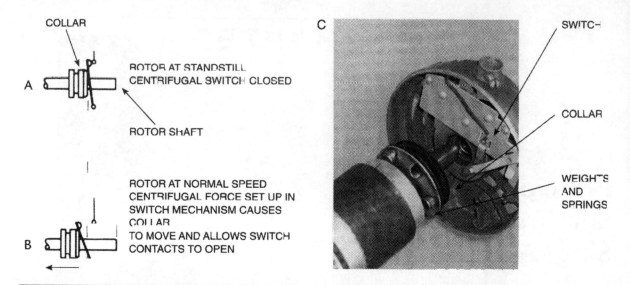

Figure 3–1 Centrifugal switch mechanism. *A.* Switch closed with no rotor movement. *B.* Switch opens to disconnect starting winding and stays open while rotor is spinning. *C.* Photo of weights, springs and activating collar; Courtesy Alerich, *Electricity 4,* Delmar Publishers.

wire. Also, because it is seated deeper in the stator iron, it has more inductive reactance. This causes the current in the running winding to lag the applied voltage by approximately 45 degrees electrically.

The second coil (starting winding) of magnet wire is wound with smaller wire so the resistance is higher than the running winding. It also has fewer turns and is placed at the top of the iron slots (inside edge) of the stator (see Figure 3–2).

Higher-resistance wire, along with less inductive reactance (less iron permeability and fewer turns), causes the current to lag the voltage by approximately 15 degrees electrically. Now the coil currents are split in the stator coils by approximately 30 degrees electrically. This is the difference between the 45-degree lag of current in the running winding and the 15-degree lag of current in the starting winding (see Figure 3–3). The relative 30-degree lag in current between the coils will produce the magnetic effects to create a rotating magnetic field.

Connections

Split-phase motors also come in dual voltage ratings. The motor connections are simple and are interchangeable from one voltage to another. If this is the case, the two sets of running windings are designed to operate in parallel with each other at 115 VAC, or in series at 230 VAC. The leads are marked with numbers and are sometimes marked with colors according to the standards set by NEMA (see Figure 3–4).

The schematic diagram for a split-phase motor helps explain the motor connections lead markings, and the method of reversing (see Figure 3–5a). If the motor is a single-voltage motor, 115 volts is applied to the main winding T^1 and T^2 and also to the starting winding. If the motor is a dual-voltage-rated motor, there are two sets of running windings. These running windings are designed to create the desired magnetic fields when they

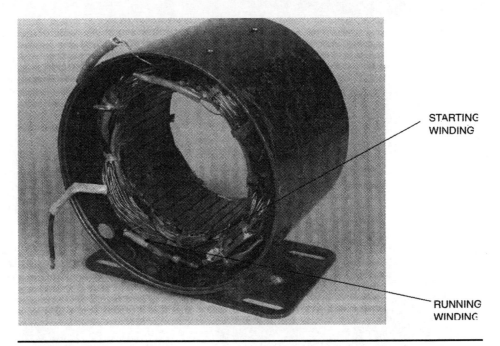

STARTING
WINDING

RUNNING
WINDING

Figure 3–2   Coils in motor show main winding deeper in slots and starting winding near the inside edge; Courtesy Keljik, *Electric Motors and Motor Controls,* Delmar Publishers.

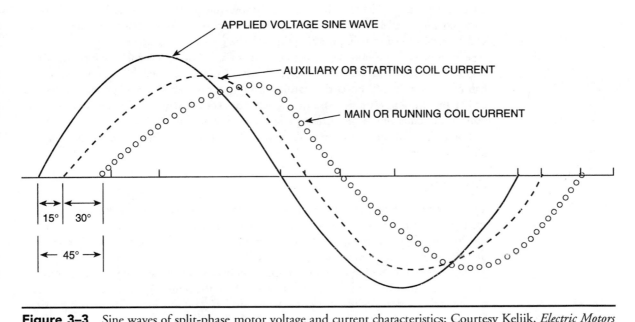

APPLIED VOLTAGE SINE WAVE

AUXILIARY OR STARTING COIL CURRENT

MAIN OR RUNNING COIL CURRENT

15°   30°

45°

Figure 3–3   Sine waves of split-phase motor voltage and current characteristics; Courtesy Keljik, *Electric Motors and Motor Controls,* Delmar Publishers.

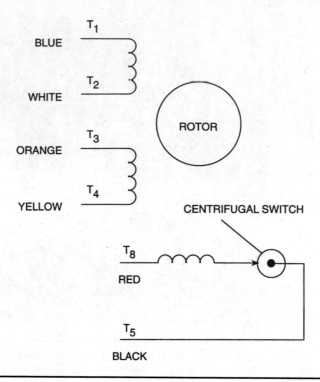

Figure 3–4   Dual-voltage, single-phase motor with lead colors. Courtesy Keljik, *Electric Motors and Motor Controls,* Delmar Publishers.

are connected in parallel to each other. For the low-voltage connection, all the windings are connected in parallel to each other. This will provide 115 volts to both set of running coils and 115 volts to the starting winding through the centrifugal switch.

Most dual-voltage motors only use one starting winding. By placing the running windings in series with each other and applying 230 volts, 115 volts will be present across each half of the winding. Connect the starting winding across the 115 volts present across one coil of the running windings (in other words, connect the starting winding from one line lead to the center of the two running winding coil connections).

Some dual-voltage motors may also have two starting windings. The connection patterns are shown in Figure 3–5b.

Reversing

To reverse the direction of rotation of the rotor, the direction of the rotating magnetic field at time of starting must be reversed. See Figure 3–6 for lead connection patterns. Reversing the direction of a dual-voltage motor running at the higher voltage, but with only one starting winding, is more complicated (see Figure 3–7).

Remember, the starting switch must open and disconnect the starting winding from the circuit or the starting winding will be damaged. Also, the centrifugal switch must be closed during locked rotor condition (in other words, rotor is not moving). If the starting switch is open, the motor will not start.

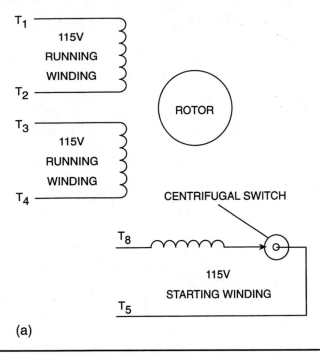

(a)

Figure 3–5a Dual-voltage, single-phase motor using one low-voltage starting winding; Courtesy Keljik, *Electric Motors and Motor Controls,* Delmar Publishers.

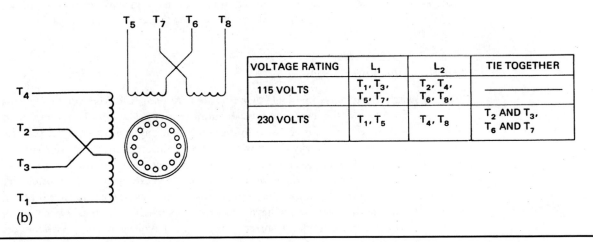

VOLTAGE RATING	L$_1$	L$_2$	TIE TOGETHER
115 VOLTS	T$_1$, T$_3$, T$_5$, T$_7$,	T$_2$, T$_4$, T$_6$, T$_8$,	——————
230 VOLTS	T$_1$, T$_5$	T$_4$, T$_8$	T$_2$ AND T$_3$, T$_6$ AND T$_7$

(b)

Figure 3–5b Connection pattern for a motor with two starting windings and two running windings; Courtesy Alerich, *Electricity 4,* Delmar Publishers.

Forward	Reverse
115V	
L_1 to: T_1, T_3, T_5, T_7	L_1 to: T_1, T_3, T_6, T_8
L_2 to: T_2, T_4, T_6, T_8	L_2 to: T_2, T_4, T_5, T_7
230V	
L_1 to: T_1, T_5	L_1 to: T_1, T_8
(connect T_2 to T_3)	(connect T_2 to T_3)
L_2 to: T_4, T_8	L_2 to: T_4, T_5
(connect T_6 to T_7)	(connect T_6 to T_7)

Figure 3–6   Forward and reverse connection patterns for dual voltage start and run windings; Courtesy Keljik, *Electric Motors and Motor Controls,* Delmar Publishers.

Forward	Reverse
115V	
L_1 to: T_1, T_3, T_5	L_1 to: T_1, T_3, T_8
L_2 to: T_2, T_4, T_8	L_2 to: T_2, T_4, T_5
230V	
L_1 to: T_1	L_1 to: T_1
(tie T_2, T_3, T_5 together)	(tie T_2, T_3, T_8 together)
L_2 to: T_4, T_8,	L_2 to: T_4, T_5

Figure 3–7   Forward and reverse connection patterns for a dual-voltage motor with a single start winding; Courtesy Keljik, *Electric Motors and Motor Controls,* Delmar Publishers.

Starting Winding Switches

Starting switches and relays for single-phase motors are used to disconnect the starting windings of a motor as the motor approaches running speed. The most common type is the centrifugal switch, which is enclosed in the motor housing and reacts to the physical speed of the rotor to centrifugally throw weights away from the shaft, causing a switch to open. See Figures 3–1 and 3–5a and b for the electrical location of the switch.

Other methods can be used where the switch is mounted externally to the motor. This may facilitate easier maintenance, or the arcing of the switch may be objectionable if placed too near the motor. In this case, a voltage (or current-operated) mechanical switch, or an electronic switch sensitive to motor current, may be used. Voltage-operated switches are connected across the starting winding (see Figure 3–8). Although these switches are used only on capacitor start motors, they will be explained here with other start switches.

When the motor is energized, the current is allowed to flow through the starting winding, which is a very low impedance circuit. However, the value of voltage dropped across the winding is too low to energize the relay or pick up the relay contacts. This is because the capacitive reactance of the capacitor is in series with the impedance of the starting winding. The large inrush current creates a large voltage drop on the capacitor and a

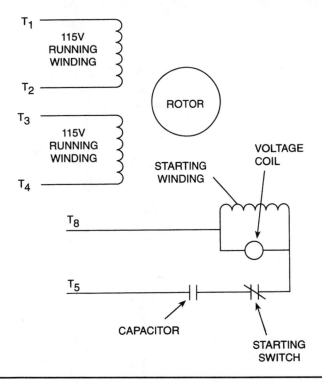

Figure 3–8 Single-phase motor using a voltage operated coil for starting circuit; Courtesy Keljik, *Electric Motors and Motor Controls,* Delmar Publishers.

lesser voltage drop on the starting winding coil. As the rotor speed increases, the voltage across the start coil increases (more voltage is induced back into the stator by the spinning rotor in the form of counterelectromotive force, which causes a higher volt drop), and the coil picks up or energizes. This opens the circuit to the start winding. The spinning rotor with its magnetic field induces a large enough voltage into the coil to hold open the contacts.

Another style of starting switch is the current-operated coil. The current-operated coil is connected in series with the run winding and the contacts are in series with the start winding (see Figure 3–9). As power is applied, the large inrush current to the run winding causes the relay to pick up. This closes the contacts to the start winding. With these contacts closed, the motor begins to spin as the speed increases, the inrush current decreases, and the current coil releases the contacts to the open position and removes the coil from the circuit.

Capacitor Start Motors

Capacitor-start motors are induction motors that range from 1/20 to 35 HP in size. These motors are commonly use for H.V.A.C. unit fans, air compressors, and saws. There are three types of capacitor-type motors: capacitor start, permanent split, and two-value capacitor motors.

Capacitor start motors are used when single-phase power is available, but there is a need for more starting torque than a split-phase motor can deliver. By using a capacitor in series with the starting winding, the relative phase angle between the starting and running winding is larger than in the split-phase motor.

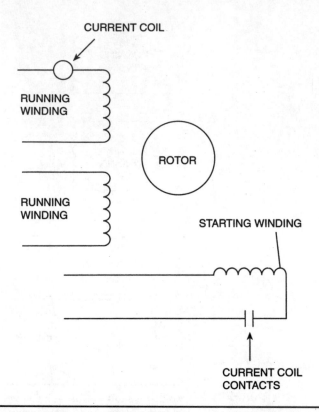

CURRENT COIL

RUNNING
WINDING

ROTOR

RUNNING
WINDING

STARTING WINDING

CURRENT COIL
CONTACTS

Figure 3–9   Single-phase motor with current operated coil used for starting winding circuit; Courtesy Keljik, *Electric Motors and Motor Controls,* Delmar Publishers.

A capacitor causes current to lead voltage, so by placing it in series with the start winding, the current in the start winding will lead the line voltage by approximately 35 degrees electrically. The run winding is similar to the split-phase motor. Its current lags the line voltage approximately 45 degrees. This corresponds to a phase split of approximately 80 degrees. This is closer to the ideal 90-degree split (see Figure 3–10).

This phase splitting accounts for the better starting torque of about 250 to 300 percent of full-load torque. It requires slightly less line current to produce the increased torque. Line current drawn is approximately two-thirds that of the same rating of a split-phase motor.

As is the case with the split-phase motor, the capacitor and starting winding are switched out of the circuit after starting. The capacitors are AC electrolytic capacitors and are not intended to be left in the circuit. They do not affect the running characteristics of the motor. Typical values of capacitors vary with HP of the motor (see Capacitor Table on next page).

Capacitors are normally mounted externally but may be mounted internally. See Figure 3–11 for an example of an externally mounted capacitor.

Connections

Capacitor start motor connections are similar to the split phase. The lead markings and the connection patterns are identical. Remember that the capacitor and the start winding are

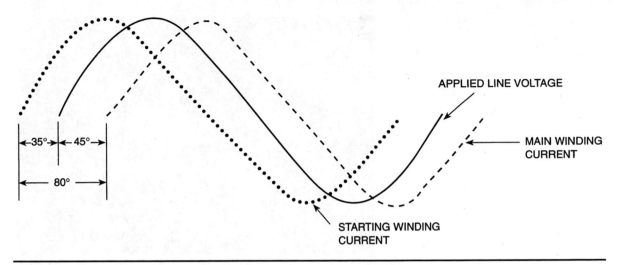

Figure 3–10 Sinewave of capacitor start motor voltage and current characteristics; Courtesy Keljik, *Electric Motors and Motor Controls,* Delmar Publishers.

HP	Microfarad
1/8	75–85
1/6	85–105
1/4	100–150
1/3	160–180
1/2	215–250
3/4	375–425
1	400–475
1 1/2	450–600
3	625–780

CAPACITOR TABLE

connected in parallel with one-half of the run windings. This is similar to the split-phase dual-voltage motor.

Reversing

Reversing the motor is accomplished the same way as the split phase. Reversing the connections to the start winding only will reverse the rotating magnetic field direction and cause the rotor to reverse, too.

Two-Value Capacitor Motors

Two-value capacitor motors employ the principle of splitting the single-phase line into two phases by using a capacitor in series with one winding. The effect is to start the motor using an electrolytic capacitor to maintain the good starting torque associated with capacitor-start induction-run motors. In fact, there is a second oil-filled running capacitor in parallel with the electrolytic. As the electrolytic capacitor is switched out of the circuit by a starting

EXTERNAL
CAPACITOR FOR
CAPACITOR START
MOTOR

Figure 3–11 Capacitor start induction-run motor; Courtesy Keljik, *Electric Motors and Motor Controls,* Delmar Publishers.

switch, the oil-filled capacitor stays in the circuit to increase the running power factor of the motor. The second effect of the running capacitor is to increase breakdown torque as the result of having a continuously rotating magnetic field in the rotor (see Figure 3–12 for a schematic of a two-value capacitor motor. See Figure 3–13a and b for sample electrolytic and oil-filled capacitors).

Permanent Split-Capacitor Motors

Motors may be constructed with a single capacitor that stays in series with one of the windings. Because this capacitor stays in the running circuit, the capacitor is an oil-filled type with a lower microfarad capacity. This lower capacity rating gives the motor less starting torque than the other two capacitor motors. It has the advantages of good running torque (similar to the two-value capacitor motor) and no starting switch that could burn out (open or shorted).

Connections

The capacitor of a permanent split capacitor motor is hooked in series with the stator windings. See Figure 3–14.

Reversing

To reverse the permanent split-capacitor motor, simply change the connection to place the capacitor in series with the other running winding (see Figure 3–14).

Testing a Capacitor

Testing a capacitor can be accomplished by performing the following:

1. Set multimeter to ohm setting.
2. Place positive probe at one end of the capacitor and the negative probe at the other end.

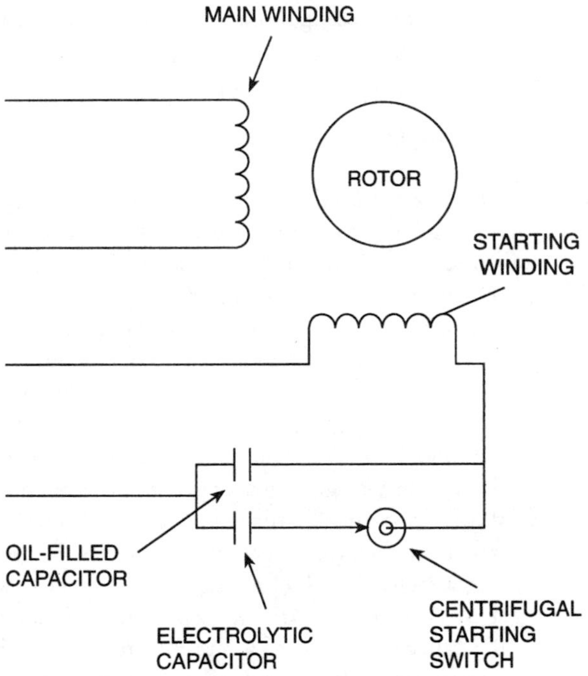
MAIN WINDING
ROTOR
STARTING WINDING
OIL-FILLED CAPACITOR
ELECTROLYTIC CAPACITOR
CENTRIFUGAL STARTING SWITCH

Figure 3–12 Schematic diagram of a two-value capacitor motor; Courtesy Keljik, *Electric Motors and Motor Controls,* Delmar Publishers.

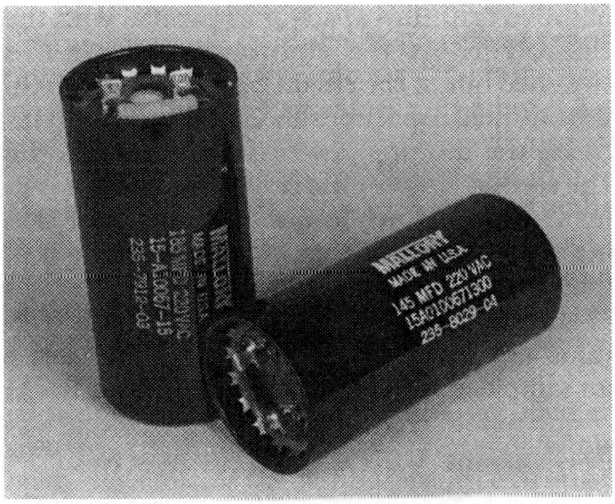

(a)

Figure 3–13a Electrolytic motor starting capacitor; Courtesy Keljik, *Electric Motors and Motor Controls,* Delmar Publishers.

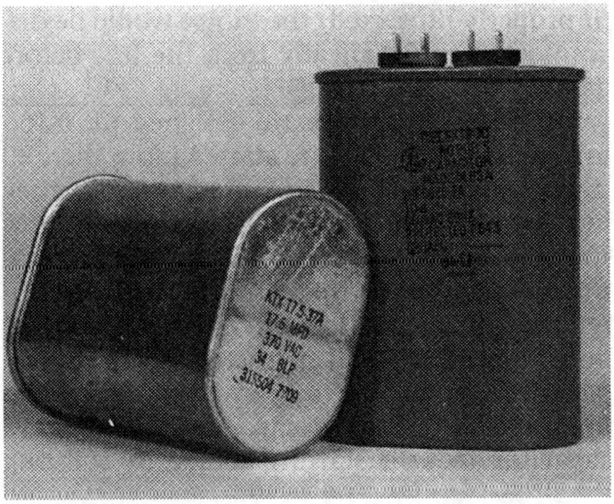

(b)

Figure 3–13b Oil-filled motor capacitor used in the running coil circuit of the motor; Courtesy Keljik, *Electric Motors and Motor Controls,* Delmar Publishers.

3. If the reading shows zero, the capacitor is bad.

4. If the reading shows infinity (pegs to the high side on an analog meter, reads high on a digital meter) and fall down slowly, the capacitor is good.

See Chapter 9 for more detailed information on testing capacitors.

Capacitor Failure and Replacement

Capacitors can fail and create an open or shorted condition. If the capacitor fails while open, the effect is the same as if the centrifugal switch is open. In this case the motor would fail to start. Do not allow the motor to stay on the line in this condition (not spinning) or the running windings will burn out. If properly protected, the motor should be disconnected automatically from the line because of the sustained starting current draw.

If the capacitor fails while shorted, the motor may fail to start because there is no longer a large phase split. It might start, but it will take a longer than normal amount of time to obtain running speed. A shorted capacitor should result in higher than normal starting current and (if properly protected) the motor would be disconnected automatically from the line before damage occurs. Starting capacitors are designed for intermittent duty. If they are started more than 20 times at three seconds each start per hour, they heat up, dry out, and fail. Therefore, capacitor-start motors are not recommended for applications that require frequent starting. Electrolytic-starting capacitors are designed for 125V operation. Too small a microfarad rating capacitor increases starting time and causes premature failure of the capacitor. Too large a capacitor reduces starting torque by splitting the phase too far. The proper capacitor should have approximately 125 percent of the line voltage developed across it during the starting cycle. The starting switch should be a snap action type that positively disconnects the capacitor from the circuit without any flutter.

Common Causes of Capacitor Failure

- Internal corrosion of the capacitor due to atmospheric conditions
- Overvoltage due to improper wiring methods, transients, starting methods, and surges
- Improper application of duty cycle, stressing the capacitor beyond its ability (overloads, worn bearings, improper-sized capacitor, low voltage, and prolonged periods of stating are key reasons for capacitor failure)
- Too-high ambient and internal temperatures

How to Properly Size the Capacitor

Often when trying to replace a capacitor, you are not able to use the old capacitor because the marking has worn off or the capacitor has exploded. The capacitor table presented earlier can be used to determine the proper size to apply. Lacking a table, you could also use the trial-and-error method.

1. Lock the rotor shaft of the motor—be careful that the shaft cannot turn.

2. Apply power to the start and run windings.

3. Record voltage across the capacitor.*

4. Measure the voltage across the start windings.

5. When the proper size capacitor is installed, the voltage reading across the capacitor should be 10 percent higher than the voltage across the start winding.

*Read voltage as quickly as possible. Values will change as the windings get hotter.

Testing the Windings Trouble- shooting

See Chapter 9 for more detailed information on testing the windings.

1. **Problem: Motor will not operate until motor shaft has been turned by hand.**

   - **Possible cause:** Motor start windings are open
   - **Possible cause:** Starting switch is open
   - **Possible cause:** Centrifugal switch is sealed in an open position
   - **Possible cause:** Open or shorted capacitor

   SOLUTION: 1. Test windings
   2. Test continuity of start switch
   3. Test capacitor
   4. Inspect the mechanical operation of the centrifugal switch

2. **Problem: Motor will operate but stalls under load or at synchronous speed.**

   - **Possible cause:** Malfunction of centrifugal switch
   - **Possible cause:** Weak or shorted capacitor
   - **Possible cause:** Bad connection at run winding
   - **Possible cause:** Open in run winding
   - **Possible cause:** Malfunction of the overload—i.e., adjustment, calibration, or damage to the overload
   - **Possible cause:** Rotor is in a bind or locked and trips the overloads

   SOLUTION: 1. Test windings
   2. Test continuity of start switch
   3. Test capacitor
   4. Inspect the mechanical operation of the centrifugal switch
   5. Test integrity and continuity of the start switch
   6. Test the continuity calibration and adjustment of the overload
   7. Inspect the start switch

3. **Problem: Motor does not start.**

   - **Possible cause:** No voltage at the start winding connections
   - **Possible cause:** Bad capacitor
   - **Possible cause:** Open in the start winding
   - **Possible cause:** Motor overloads are open

   SOLUTION: 1. Test windings
   2. Test continuity of start switch
   3. Test capacitor
   4. Inspect the mechanical operation of the centrifugal switch
   5. Test integrity and continuity of the start switch
   6. Test the continuity calibration and adjustment of the overload
   7. Inspect the start switch

CHAPTER

4

Repulsion Single-Phase Motors*

**Repulsion
Motor
Construction**

A repulsion motor has the following parts:

- **Laminated stator core.** This winding is similar to the main or running winding of a split-phase motor. The stator usually is wound with four, six, or eight poles.
- **Rotor consisting of a slotted core into which a winding is placed.** The rotor is similar in construction to the armature of a DC motor. Thus, the rotor is called an armature. The coils that make up this armature winding are connected to a commutator. The commutator has segments or bars parallel to the armature shaft.
- **Carbon brushes contacting with the commutator surface.** The brushes are held in place by a brush holder assembly mounted on one of the end shields. The brushes are connected together by heavy copper jumpers. The brush holder assembly may be moved so that the brushes can make contact with the commutator surface at different points to obtain the correct rotation and maximum torque output. There are two types of brush arrangements:
  1. Brush riding—the brushes are in contact with the commutator surface at all times.
  2. Brush lifting—the brushes lift at approximately 75 percent of the rotor speed.
- **Two cast steel-end shields.** These shields house the motor bearings and are secured to the motor frame.
- **Two bearings supporting the armature shaft.** The bearings center the armature with respect to the stator core and windings. The bearings may be sleeve bearings or ball-bearing units.
- **Cast steel frame into which the stator core is pressed**

**Operation
of a Repulsion
Motor**

The connection of the stator winding of a repulsion motor to a single-phase line causes a field to be developed by the current in the stator windings. This stator field induces a voltage and a resultant current in the rotor windings. If the brushes are placed in the proper

*Written with permission from Alerich, *Electricity 4,* Delmar Publishers.

position on the commutator segments, the current in the armature windings will set up proper magnetic poles in the armature.

These armature field poles have a set relationship to the stator field poles. That is, the magnetic poles developed in the armature are set off from the field poles of the stator winding by about 15 electrical degrees. Furthermore, since the instantaneous polarity of the rotor poles is the same as that of the adjacent stator poles, the repulsion torque created causes the rotation of the motor armature.

The three diagrams of Figure 4–1 show the importance of the brushes being in the proper position to develop maximum torque. In Figure 4–1a, no torque is developed when the brushes are placed at right angles to the stator poles. This is due to the fact that the equal-induced voltages in the two halves of the armature winding oppose each other at the connection between the two sets of brushes. Since there is no current in the windings, flux is not developed by the armature windings.

In Figure 4–1b, the brushes are in a position directly under the center of the stator poles. A heavy current exists in the armature windings with the brushes in this position, but there is still no torque. The heavy current in the armature windings sets up poles in the armature. However, these poles are centered with the stator poles and a torque is not created, either in a clockwise or counterclockwise direction.

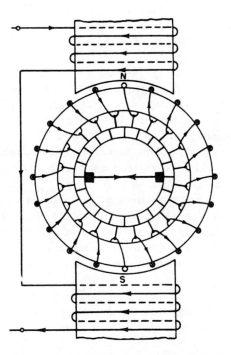 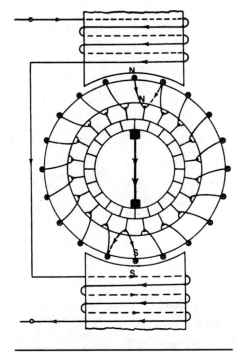

Figure 4–1a No torque created, equal voltage values oppose each other; Courtesy Alerich, *Electricity 4,* Delmar Publishers.

Figure 4–1b No torque, even though current value in armature is high; Courtesy Alerich, *Electricity 4,* Delmar Publishers.

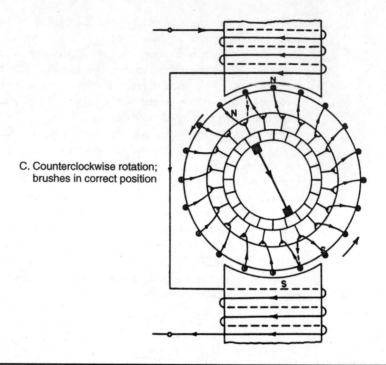

C. Counterclockwise rotation;
brushes in correct position

Figure 4–1c Repulsion motor operation; Courtesy Alerich, *Electricity 4*, Delmar Publishers.

In Figure 4–lc, the brushes have shifted from the center of the stator poles 15 electrical degrees in a counterclockwise direction. Thus, magnetic poles of like polarity are set up in the armature. These poles are 15 electrical degrees in a counterclockwise direction from the stator pole centers. A repulsion torque is created between the stator and the rotor field poles of like polarity. The torque causes the armature to rotate in a counterclockwise direction. A repulsion machine has a high starting torque, with a small starting current, and a rapidly decreasing speed with an increasing load.

The direction of rotation of a repulsion motor is reversed if the brushes are shifted 15 electrical degrees from the stator field pole centers in a clockwise direction, as in Figure 4–2. As a result, magnetic poles of like polarity are set up in the armature. These poles are 15 electrical degrees in a clockwise direction from the stator pole centers. Repulsion motors are used principally for constant-torque applications, such as printing-press drives, fans, and blowers.

Repulsion Start, Induction-Run Motor

A second type of repulsion motor is the repulsion start, induction-run motor. In this type of motor, the brushes contact the commutator at all times. The commutator of this motor is the more conventional axial form.

A repulsion start, induction-run motor consists basically of the following parts.

- **Laminated stator core.** This core has one winding, which is similar to the main or running winding of a split-phase motor.
- **Rotor consisting of a slotted core into which a winding is placed.** The coils that make up the winding are connected to a commutator. The rotor core and winding are similar to the armature of a DC motor. Thus, the rotor is called an armature.

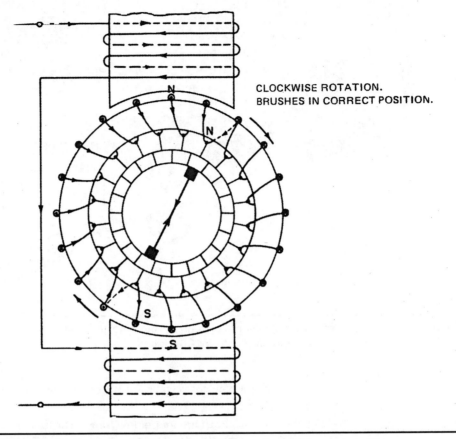

CLOCKWISE ROTATION.
BRUSHES IN CORRECT POSITION.

Figure 4–2 Reversing the direction of rotation of a repulsion motor; Courtesy Alerich, *Electricity 4*, Delmar Publishers.

- **Centrifugal device.** In the brush-lifting type of motor, there is a centrifugal device that lifts the brushes from the commutator surface at 75 percent of the rated speed. This device consists of governor weights, a short-circuiting necklace, a spring barrel, spring, push rods, brush holders, and brushes. Although high in first cost, this device does save wear and tear on brushes, and runs quietly. Figure 4–3 is an exploded view of the armature, radial commutator, and centrifugal device of the brush-lifting type of repulsion start, induction-run motor.

   The brush-riding type of motor also contains a centrifugal device that operates at 75 percent of the rated speed. This device consists of governor weights, a short-circuiting necklace, and a spring barrel. The commutator segments are short circuited by this device, but the brushes and brush holders are not lifted from the commutator surface.

- **Commutator.** The brush-lifting type of motor has a radial-type commutator (Figure 4–3). The brush-riding type of motor has an axial commutator (Figure 4–4).

- **Brush holder assembly.** The brush holder assembly for the brush-lifting type of motor is arranged so that the centrifugal device can lift the brush holders and brushes clear of the commutator surface.

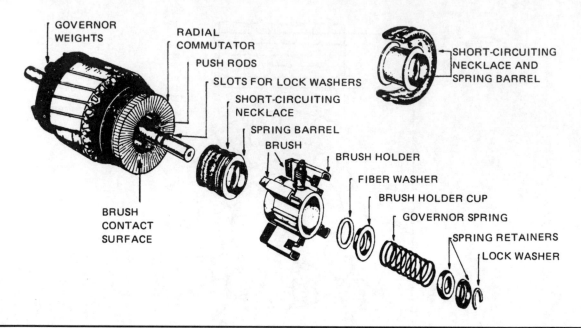

Figure 4–3   An exploded view of a radial commutator and centrifugal brush-lifting device for a repulsion start, induction-run motor; Courtesy Alerich, *Electricity 4,* Delmar Publishers.

The brush holder assembly for the brush-riding type of motor is the same as that of a repulsion motor.

- **End shields, bearings, and motor frame.** The parts are the same as those of a repulsion motor.

Operation of the Centrifugal Mechanism

Refer to Figure 4–3 to identify the components of the centrifugal mechanism.

The operation of this device consists of the following steps. As the push rods of the centrifugal device move forward, they push the spring barrel forward. This allows the short-circuiting necklace to make contact with the radial commutator bars, which thus are all short circuited. At the same time, the brush holders and brushes are moved from the commutator surface. As a result, there is no unnecessary wear on the brushes and the commutator surface and there are no objectionable noises caused by the brushes riding on the radial commutator surface.

The short-circuiting action of the governor mechanism and the commutator segments converts the armature to a form of squirrel-cage rotor and the motor operates as a single-phase induction motor. In other words, the motor starts as a repulsion motor and runs as an induction motor.

In the brush-riding type of motor, an axial commutator is used. The centrifugal mechanism (Figure 4–4) consists of a number of copper segments that are held in place by a spring. This device is placed next to the commutator. When the rotor reaches 75 percent of the rated speed, the centrifugal device short circuits the commutator segments. The motor then will continue to operate as an induction motor.

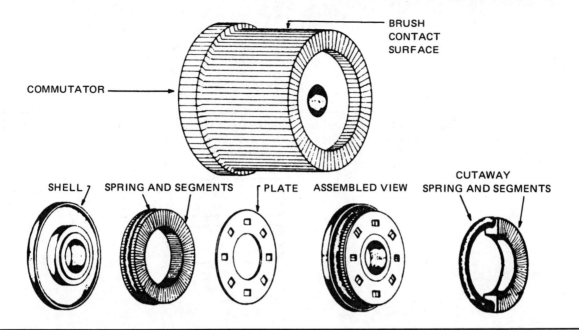

BRUSH
CONTACT
SURFACE

COMMUTATOR

SHELL   SPRING AND SEGMENTS   PLATE   ASSEMBLED VIEW   CUTAWAY
SPRING AND SEGMENTS

Figure 4–4   An exploded view of a short-circuiting device for a brush-riding, repulsion start, induction-run motor; Courtesy Alerich, *Electricity 4,* Delmar Publishers.

Operation of a Repulsion Start, Induction-Run Motor

The starting torque is good for either the brush-lifting type or the brush-riding type of repulsion start, induction-run motor. Furthermore, the speed performance of both types of motors is very good, since they operate as single-phase induction motors.

Because of the excellent starting and running characteristics for both types of repulsion start, induction-run motors, they are used for a variety of industrial applications, including commercial refrigerators, compressors, and pumps.

The direction of rotation for a repulsion start, induction-run motor is changed in the same manner as that for a repulsion motor—that is, by shifting the brushes past the stator pole center 15 electrical degrees. The symbol in Figure 4–5 represents both a repulsion start, induction run motor and a repulsion motor.

Many repulsion start, induction-run motors are designed to operate on 115 volts or 230 volts. These dual-voltage motors contain two stator windings. For 115-volt operation, the stator windings are connected in parallel; for 230-volt operation, the stator windings are connected in series. The diagram and photo in Figure 4–6 present a dual-voltage, repulsion start, induction-run motor. The connection table in Figure 4–6 shows how the leads of the motor are connected for either 115-volt operation or 230-volt operation. These connections also can be used for dual-voltage repulsion motors.

Repulsion-Induction Motor

The operating characteristics of a repulsion-induction motor are similar to those of the repulsion start, induction-run motor. However, the repulsion-induction motor has no centrifugal mechanism. It has the same type of armature and commutator as the repulsion motor, but it has a squirrel-cage winding beneath the slots of the armature.

Figure 4–5 Schematic diagram symbol of a repulsion start, induction run motor and a repulsion motor; Courtesy Alerich, *Electricity 4,* Delmar Publishers.

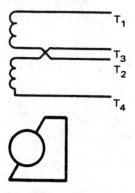

	L₁	L₂	TIE TOGETHER
LOW VOLTAGE	T₁ T₃	T₂ T₄	————
HIGH VOLTAGE	T₁	T₄	T₂ T₃

Figure 4–6 Schematic diagram of a dual-voltage, repulsion start, induction run motor; Courtesy Alerich, *Electricity 4,* Delmar Publishers.

Figure 4–7 shows a repulsion-induction motor armature with a squirrel-cage winding. One advantage of this type of motor is that it has no centrifugal device requiring maintenance. The repulsion-induction motor has a very good starting torque since it starts as a repulsion motor. At start up, the repulsion winding predominates; but, as the motor speed increases, the squirrel-cage winding becomes predominant. The transition from repulsion to induction operation is smooth, since no switching device is used. In addition, the repulsion-induction motor has a fairly constant speed regulation, from no load to full load, because of the squirrel-cage winding. The torque-speed performance of a repulsion-induction motor is similar to that of a DC compound motor.

A repulsion-induction motor can be operated on either 115 volts or 230 volts. The stator winding has two sections which are connected in parallel for 115-volt operation, and in series for 230-volt operation. The markings of the motor terminals and the connection arrangement of the leads is the same as in a repulsion start, induction-run motor.

The symbol in Figure 4–5 also represents a repulsion-induction motor (as well as a repulsion start, induction-run motor and a repulsion motor.)

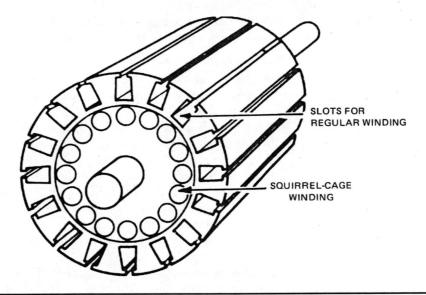

SLOTS FOR
REGULAR WINDING

SQUIRREL-CAGE
WINDING

Figure 4–7 An armature of a repulsion-induction motor; Courtesy Alerich, *Electricity 4*, Delmar Publishers.

Reversing

Reversing a repulsion-type motor is accomplished by using two sets of windings, 90 electrical degrees apart, in the stator. These two windings are hooked in series to one another. Reversing shaft rotation is accomplished by reversing the leads of these two windings.

Trouble-shooting

1. **Problem: Motor will not start, centrifugal switch is closed.**
   - **Possible cause:** Open overcurrent protection device such as blown fuse or tripped circuit breaker. Replace and find out what caused the device to trip. An overcurrent device usually trips for one of the following reasons: ground fault, short circuit, overload, or damaged device.
   - **Possible cause:** Tripped overload—check the overload adjustment if applicable. Perform amp draw test at motor or starter. Replace or readjust the overload if needed.
   - **Possible cause:** Wrong brush position or worn brushes. When brushes do not contact commutator, the motor will not start. Check for a dirty commutator or poor spring tension at brush holder.
   - **Possible cause:** Loose shaft, worn or damaged bearings. Bearings keep the rotor from touching the stator winding. If the rotor touches the stator windings a short circuit will occur due to the wear of the turning action. Test the bearings by moving the shaft vertically. If the shaft moves vertically the bearings are worn and should be replaced. When replacing the bearings, check the armature and stator windings for worn or bare places. Repair all damaged surfaces of the armature and stator.
   - **Possible cause:** Dirty commutator—Clean with a penciling tool or other acceptable means such as emery cloth. An isolating glaze develops on the commutator with the presence of dust, other similar particles, and prolonged use.
   - **Possible cause:** Open, bad, or wrong connections at windings or brushes. Often the necklace welds to the commutator and grounds out the commutator bars.

- **Possible cause:** Damaged necklace assembly shorts to the armature. Often the necklace welds to the commutator and grounds out the commutator bars.

2. **Problem: Motor gets too hot.**

- **Possible cause:** Ambient temperature is higher than the rating of motor—See temperature rating at nameplate. The motor must not be subjected to higher temperature than its rating.
- **Possible cause:** Loose shaft, worn or damaged bearings. Bearings keep the rotor from touching the stator winding. If the rotor touches the stator windings a short circuit will occur due to the wear of the turning action. Test the bearings by moving the shaft vertically. If the shaft moves vertically, the bearings are worn and should be replaced. When replacing the bearings, check the armature and stator windings for worn or bare places. Repair all damaged surfaces of the armature and stator.
- **Possible cause:** Short in armature or stator windings—See testing section at the end of this chapter.
- **Possible cause:** Improper brush placement. When brushes do not contact commutator, the motor will not start. Check for a dirty commutator, damaged brush holder, or poor spring tension at brush holder.
- **Possible cause:** Damaged necklace assembly. Often the necklace welds to the commutator and grounds out the commutator bars.
- **Possible cause:** Low voltage.

3. **Problem: Motor becomes noisy.**

- **Possible cause:** Loose shaft, worn or damaged bearings. Bearings keep the rotor from touching the stator winding. If the rotor touches the stator windings, a short circuit will occur due to the wear of the turning action. Test the bearings by moving the shaft vertically. If the shaft moves vertically, the bearings are worn and should be replaced. When replacing the bearings, check the armature and stator windings for worn or bare places. Repair all damaged surfaces of the armature and stator.
- **Possible cause:** Loose centrifugal assembly. See section on centrifugal switching.
- **Possible cause:** Short in the stator windings. See testing section of this handbook.

4. **Problem: Overcurrent protection device is tripped.**

- **Possible cause:** Ground fault or short circuit in field windings. See testing section.
- **Possible cause:** Improper lead connections.
- **Possible cause:** Short circuit or ground fault in the armature.
- **Possible cause:** Brushes improperly set or not touching the commutator. When brushes do not contact commutator, the motor will not start. Check for a dirty commutator, damaged brush holder, or poor spring tension at brush holder.
- **Possible cause:** Bearings are locked up. The shaft cannot turn.

5. **Problem: The motor hums at start up and does not operate.**

- **Possible cause:** Improper lead connections to the brushes and stator winding.
- **Possible cause:** Loose, worn, or damaged bearings. Bearings keep the rotor from touching the stator winding. If the rotor touches the stator windings, a short

circuit will occur due to the wear of the turning action. Test the bearings by moving the shaft vertically. If the shaft moves vertically, the bearings are worn and should be replaced. When replacing the bearings, check the armature and stator windings for worn or bare places. Repair all damaged surfaces of the armature and stator.

- **Possible cause:** Improper brush position. When brushes do not contact commutator, the motor will not start. Check for a dirty commutator, damaged brush holder, or poor spring tension at brush holder.
- **Possible cause:** Short circuit or ground fault in armature and stator windings
- **Possible cause:** Brushes not touching the commutator
- **Possible cause:** Dirty or damaged commutator—Clean with a penciling tool or other acceptable means such as emery cloth. An isolating glaze develops on the commutator with the presence of dust, other similar particles, and prolonged use.

6. **Problem: Motor will not come up to speed.**

- **Possible cause:** Improper brush contact with the commutator. When brushes do not contact commutator, the motor will not start. Check for a dirty commutator, damaged brush holder, or poor spring tension at brush holder.
- **Possible cause:** Fault in the necklace assembly. Often the necklace welds to the commutator and grounds out the commutator bars.
- **Possible cause:** Dirty or damaged commutator. Clean with a penciling tool or other acceptable means such as emery cloth. An isolating glaze develops on the commutator with the presence of dust, other similar particles, and prolonged use.
- **Possible cause:** Loose, worn, or bad bearings. Bearings keep the rotor from touching the stator winding. If the rotor touches the stator windings a short circuit will occur due to the wear of the turning action. Test the bearings by moving the shaft vertically. If the shaft moves vertically, the bearings are worn and should be replaced. When replacing the bearings, check the armature and stator windings for worn or bare places. Repair all damaged surfaces of the armature and stator.
- **Possible cause:** Brush holder is held too far from the commutator because push rods are too long. The brush holder should be at about 1/32 inch from the commutator when staring.

7. **Problem: Arcing and sparking is noticed while motor is running.**

- **Possible cause:** Armature coil is open.
- **Possible cause:** Dirty commutator. Clean with a penciling tool or other acceptable means such as emery cloth. An isolating glaze develops on the commutator with the presence of dust, other similar particles, and prolonged use.
- **Possible cause:** Improper brush contact on commutator.
- **Possible cause:** High mica. When the copper bars of the armature wear down below the mica between the bars, the mica keeps the brush from making contact with the commutator.

Three-Phase Motors*

Three-Phase, Squirrel-Cage Induction Motor

Operating Characteristics

The three-phase, squirrel-cage induction motor is relatively small in physical size or a given horsepower rating when compared with other types of motors. The squirrel-cage induction motor has very good speed regulation under varying load conditions.

Because of its rugged construction and reliable operation, the three-phase, squirrel-cage induction motor is widely used for many industrial applications.

Construction Details

The three-phase, squirrel-cage induction motor normally consists of a stator, a rotor, and two end shields housing the bearings that support the rotor shaft.

A minimum of maintenance is required with this type of motor because:

- the rotor has no windings to become shorted
- there are no commutator or slip rings to service (compared to the DC motor)
- there are no brushes to replace

The motor frame is usually made of cast steel. The stator core is pressed directly onto the frame. The two end shields housing the bearings are bolted to the cast steel frame. The bearings that support the rotor shaft are either sleeve bearings or ball bearings.

Stator A typical stator contains a three-phase winding mounted in the slots of a laminated steel core. The winding itself consists of formed coils of wire connected so that there are three single-phase windings spaced 120 electrical degrees apart. The three separate single-phase windings are then connected, usually internally, in either wye or delta. Three or nine leads from the three-phase stator windings are brought out to a terminal box mounted on the frame of the motor for single- or dual-voltage connections.

Rotor The revolving part of the motor consists of steel punchings or laminations arranged in a cylindrical core. Copper or aluminum bars are mounted near the surface of

*Written with permission from Alerich, *Electricity 4*, Delmar Publishers.

the rotor. The bars are brazed or welded to two copper end rings. In some small squirrel-cage induction motors, the bars and end rings are cast in one piece from aluminum.

Note that fins are cast into the rotor to circulate air and cool the motor while it is running. Note also that the rotor bars between the windings are at an angle to the faces of the rings. Because of this design, the running motor will be quieter and smoother in operation. A keyway is visible on the left end of the shaft. A pulley or load shaft coupling can be secured using this keyway.

Shaft Bearings   The inside walls of the sleeve bearings are made of a babbitt metal that provides a smooth, polished, and long-wearing surface for the rotor shaft. An oversized oil ring fits loosely around the rotor shaft and extends down into the oil reservoir. This ring picks up and slings oil on the rotating shaft and bearing surfaces. Oil-ring lubrication minimizes friction losses. An oil-inspection cup on the side of each end shield enables maintenance personnel to check the level of the oil in the sleeve bearing.

In some motors, ball bearings are used instead of sleeve bearings. Grease, rather than oil, is used to lubricate ball bearings. This type of bearing usually is two-thirds full of grease at the time the motor is assembled. Special fittings are provided on the end bells so that a grease gun can be used to apply additional lubricant to the ball-bearing units at periodic intervals.

When lubricating roller bearings, remove the bottom plug so that the old grease is forced out. The manufacturer's specifications for the motor should be consulted for the lubricant grade recommended, the lubrication procedure, and the bearing loads.

Principle of Operation of a Squirrel-Cage Motor

As stated in a previous paragraph on the stator construction, the slots of the stator core contain three separate single-phase windings. When three currents 120 electrical degrees apart pass through these windings, a rotating magnetic field results. This field travels around the inside of the stator core. The speed of the rotating magnetic field depends on the number of stator poles and the frequency of the power source. This speed is called the synchronous speed and is determined by the formula:

$$S = \frac{120 \times f}{p}$$

Where:

S = Synchronous speed
f = Hertz (frequency)
p = Number of poles

Example 1   If a three-phase, squirrel-cage induction motor has six poles on the stator winding and is connected to a three-phase, 60-hertz source, then the synchronous speed of the revolving field is 1,200 r/min.

$$S = \frac{120 \times f}{p} = \frac{120 \times 60}{6} = 1,200 \text{ r/min}$$

As this magnetic field rotates at synchronous speed, it cuts the copper bars of the rotor and induces voltages in the bars of the squirrel-cage winding. These induced voltages set up currents in the rotor bars, which, in turn, create a field in the rotor core. This rotor field reacts with the stator field to cause a twisting effect or torque, which turns the rotor. The rotor always turns at a speed slightly less than the synchronous speed of the stator field. This means that the stator field will cut the rotor bars. If the rotor turns at the same speed as the stator field, the stator field will not cut the rotor bars and there will be no torque.

Speed Regulation and Percent Slip   The squirrel-cage induction motor has very good speed regulation characteristics (the ratio of difference in speed from no load to full load). Speed performance is measured in terms of percent slip. The synchronous speed of the rotating field of the stator is used as a reference point. Recall that the synchronous speed depends on the number of stator poles and the operating frequency. Since these two quantities remain constant, the synchronous speed also remains constant. If the speed of the rotor at full load is deducted from the synchronous speed of the stator field, the difference is the number of revolutions per minute that the rotor slips behind the rotating field of the stator.

$$\text{Percent Slip} = \frac{\text{Synchronous speed} - \text{rotor speed}}{\text{Synchronous speed}} \times 100$$

Example 2   If the three-phase, squirrel-cage induction motor used in Example 1 has a synchronous speed of 1,200 r/min and a full-load speed of 1,140 r/min, find the percent of slip.

$$\text{Synchronous speed (Example 1)} = 1,200 \text{ r/min}$$

$$\text{Full-load rotor speed} = 1,140 \text{ r/min}$$

$$\text{Percent slip} = \frac{\text{Synchronous speed} - \text{rotor speed}}{\text{Synchronous speed}} \times 100$$

$$\text{Percent slip} = \frac{1,200 - 1,140}{1,200} \times 100$$

$$\text{Percent slip} = \frac{60 \times 100}{1,200} = .05 \times 100$$

$$\text{Percent slip} = 5 \text{ percent}$$

For a squirrel-cage induction motor, as the value of percent slip decreases, the speed performance of the motor is improved. The average range of percent slip for squirrel-cage induction motors is 2 percent to 6 percent.

Figure 5–1 shows a speed curve and a percent slip curve for a squirrel-cage induction motor operating between no load and full load. The rotor speed at no load slips behind the synchronous speed of the rotating stator field just enough to create the torque required to overcome friction and windage losses at no load. As a mechanical load is applied to the motor shaft, the rotor tends to slow down. This means that the stator field (turning at a fixed speed) cuts the rotor bars a greater number of times in a given period. The induced voltages in the rotor bars increase, resulting in more current in the rotor bars and a stronger rotor field. There is a greater magnetic reaction between the stator and rotor fields, which causes a stronger twisting effect or torque. This also increases stator current taken from the line. The motor is able to handle the increased mechanical load with very little decrease in the speed of the rotor.

Typical slip-torque curves for a squirrel-cage induction motor are shown in Figure 5–2. The torque output of the motor in foot-pound (ft lb) increases as a straight line with an increase in the value of percent slip as the mechanical load is increased to the point of full load. Beyond full load, the torque curve bends and finally reaches a maximum point called the breakdown torque. If the motor is loaded beyond this point, there will be a corresponding decrease in torque until the point is reached where the motor stalls. However, all induction motors have some slip in order to function.

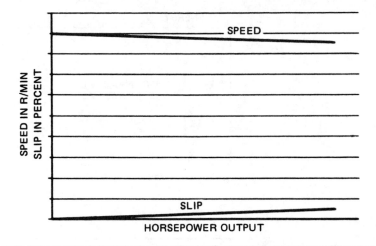

Figure 5–1 Speed and percentage slip curve; Courtesy Alerich, *Electricity 4,* Delmar Publishers.

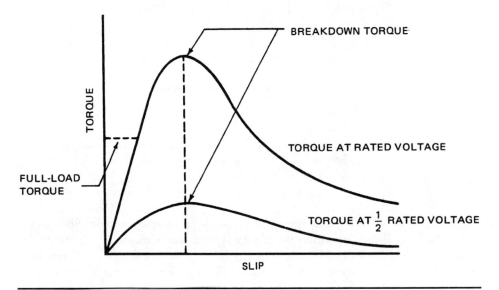

Figure 5–2 Slip-torque curves for a squirrel-cage motor; Courtesy Alerich, *Electricity 4,* Delmar Publishers.

Starting Current When a three-phase, squirrel-cage induction motor is connected across the full line voltage, the starting surge of current momentarily reaches as high a value as 400 to 600 percent of the rated full-load current. At the moment the motor starts, the rotor is at a standstill. At this instant, therefore, the stator field cuts the rotor bars at a faster rate than when the rotor is turning. This means that there will be relatively high induced voltages in the rotor, which will cause heavy rotor currents. The resulting input current to the stator windings will be high at the instant of starting. Because of this high starting current, starting protection rated as high as 300 percent of the rated full-load current is provided for squirrel-cage induction motor installations.

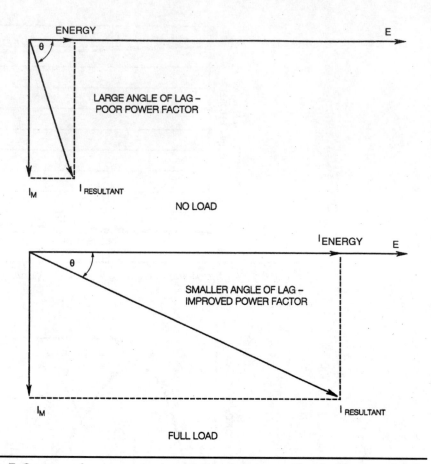

Figure 5–3 Power factor at no load and full load; Courtesy Alerich, *Electricity 4,* Delmar Publishers.

Most squirrel-cage induction motors are started at full voltage. If there are any questions concerning the starting of large sizes of motors at full voltage, the electric utility company should be consulted. In the event that the feeders and protective devices of the electric utility are unable to handle the large starting currents, reduced-voltage starting circuits must be used with the motor.

Power Factor   The power factor of a squirrel-cage induction motor is poor at no-load and low-load conditions. At no load, the power factor can be as low as 15 percent lagging. However, as load is applied to the motor, the power factor increases. At the rated load, the power factor may be as high as 85 to 90 percent lagging.

The power factor at no load is low because the magnetizing component of input current is a large part of the total input current of the motor. When the load on the motor is increased, the in-phase current supplied to the motor increases, but the magnetizing component of current remains practically the same. This means that the resultant line current is more nearly in phase with the voltage and the power factor is improved when the motor is loaded, compared with an unloaded motor, which draws its magnetizing current chiefly.

Figure 5–3 shows the increase in power factor from a no-load condition to full load. In the no-load diagram, the in-phase current (*I* energy) is small when compared to the mag-

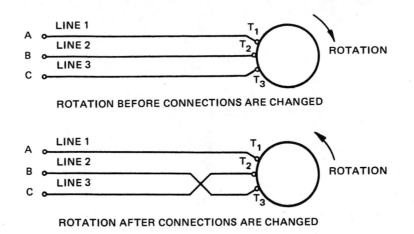

Figure 5–4   Reversing rotation of an induction motor; Courtesy Alerich, *Electricity 4,* Delmar Publishers.

netizing current (I^M); thus, the power factor is poor at no load. In the full-load diagram, the in-phase current has increased while the magnetizing current remains the same. As a result, the angle of lag of the line current decreases and the power factor increases.

Reversing Rotation   The direction of rotation of a three-phase induction motor can be reversed readily. The motor will rotate in the opposite direction if any two of the three line leads are reversed (Figure 5–4). The leads are reversed at the motor.

Speed Control   A squirrel-cage induction motor has almost no speed control. Recall that the speed of the motor depends on the frequency of the three-phase source and the number of poles of the stator winding.

The frequency of the supply line is usually 60 hertz, and is maintained at this value by the local power utility company. Since the number of poles in the motor is also a fixed value, the synchronous speed of the motor remains constant. As a result, it is not possible to obtain a range of speed. The three-phase, squirrel-cage induction motor, therefore, is used for applications where a wide range of speed is not necessary.

Induction Motors with Dual-Voltage Connections

Many three-phase, squirrel-cage induction motors are designed to operate at two different voltage ratings. For example, a typical dual-voltage rating for a three-phase motor is 230/460 volts.

Figure 5–5 shows a typical wye-connected stator winding that may be used for either 230 volts, three phase or 460 volts, three phase. Each of the three single-phase windings consists of two coil windings. There are nine leads brought out externally from this type of stator winding. These leads, identified as leads 1 to 9, end in the terminal box of the motor. To mark the terminals, start at the upper left-hand terminal T^1 and proceed in a clockwise direction in a spiral toward the center, marking each lead as indicated in the figure.

Figure 5–6 shows the connections required to operate a motor from a 460-volt, three-phase source. The two coils of each single-phase winding are connected in series. Figure 5–7 shows the connections to permit operation from a 230-volt, three-phase source.

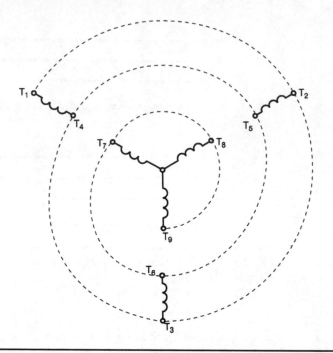

Figure 5–5 Method of identifying terminal markings; Courtesy Alerich, *Electricity 4,* Delmar Publishers.

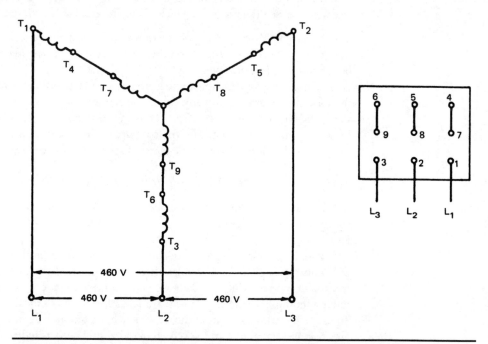

Figure 5–6 460 volt wye connection—coils connected in series; Courtesy Alerich, *Electricity 4,* Delmar Publishers.

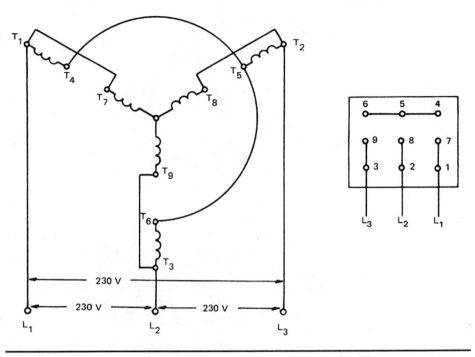

Figure 5–7 230 volt wye connection—coils connected in series; Courtesy Alerich, *Electricity 4,* Delmar Publishers.

Three-Phase, Wound-Rotor Motor

Many industrial motor applications require three-phase motors with variable speed control. The squirrel-cage induction motor cannot be used for variable speed work since its synchronous speed is essentially constant. Therefore, another type of induction motor was developed for variable speed applications. This motor is called the wound-rotor induction motor or slip-ring AC motor.

Construction Details

A three-phase, wound-rotor induction motor consists of a stator core with a three-phase winding, a wound rotor with slip rings, brushes and brush holders, and two endshields to house the bearings that support tie rotor shaft. Figure 5–8 shows the basic parts of a three-phase, wound-rotor induction motor.

The Stator A typical stator contains a three-phase winding held in place in the slots of a laminated steel core. The winding consists of formed coils arranged and connected so that there are three single-phase windings spaced 120 electrical degrees apart. The separate single-phase windings are connected either in wye or delta. Three line leads are brought out to a terminal box mounted on the frame of the motor. This is the same construction as the squirrel-cage motor stator.

The Rotor The rotor consists of a cylindrical core composed of steel laminations. Slots cut into the cylindrical core hold the formed coils of wire for the rotor winding.

The rotor winding consists of three single-phase windings spaced 120 electrical degrees apart. The single-phase windings are connected either in wye or delta. (The rotor winding

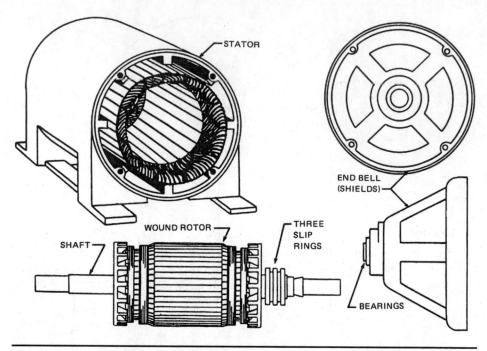

Figure 5–8 Parts of a wound-rotor motor; Courtesy Alerich, *Electricity 4,* Delmar Publishers.

must have the same number of poles as the stator winding.) The three leads from the three-phase rotor winding terminate at three slip rings mounted on the rotor shaft. Leads from carbon brushes that ride on these slip rings are connected to an external speed controller to vary the rotor resistance for speed control.

The brushes are held securely to the slip rings of the wound rotor by adjustable springs mounted in the brush holders. The brush holders are fixed in one position. For this type of motor, it is not necessary to shift the brush position, as is sometimes required in direct-current generator and motor work.

The Motor Frame The motor frame is made of cast steel. The stator core is pressed directly into the frame. Two end shields are bolted to the cast steel frame. One of the end shields is larger than the other because it must house the brush holders and brushes that ride on the slip rings of the wound rotor. In addition, it often contains removable inspection covers.

The bearing arrangement is the same as that used in squirrel-cage induction motors. Either sleeve bearings or ball-bearing units are used in the end shields.

Principle of Operation

When three currents, 120 electrical degrees apart, pass through the three single-phase windings in the slots of the stator core, a rotating magnetic field is developed. This field travels around the stator. The speed of the rotating field depends on the number of stator poles and the frequency of the power source. This speed is called the synchronous speed. It is determined by applying the formula that was used to find the synchronous speed of the rotating field of squirrel-cage induction motors:

$$S = \frac{120 \times f}{p}$$

As the rotating field travels at synchronous speed, it cuts the three-phase winding of the rotor and induces voltages in this winding. The rotor winding is connected to the three slip rings mounted on the rotor shaft. The brushes riding on the slip rings connect to an external wye-connected group of resistors (speed controller).

The induced voltages in the rotor windings set up currents that follow a closed path from the rotor winding to the wye-connected speed controller. The rotor currents create a magnetic field in the rotor core based on transformer action. This rotor field reacts with the stator field to develop the torque, which causes the rotor to turn. The speed controller is sometimes called the secondary resistance control.

Starting Theory of Wound-Rotor Induction Motors

To start the motor, all of the resistance of the wye-connected speed controller is inserted in the rotor circuit. The stator circuit is energized from the three-phase line. The voltage induced in the rotor develops currents in the rotor circuit. The rotor currents, however, are limited in value by the resistance of the speed controller. As a result, the stator current also is limited in value. In other words, to minimize the starting surge of current to a wound-rotor induction motor, insert the full resistance of the speed controller in the rotor circuit. As the motor accelerates, steps of resistance in the wye-connected speed controller can be cut out of the rotor circuit until the motor accelerates to its rated speed.

Speed Control   The insertion of resistance in the rotor circuit not only limits the starting surge of current, but also produces a high starting torque and provides a means of adjusting the speed. If the full resistance of the speed controller is cut into the rotor circuit when the motor is running, the rotor current decreases and the motor slows down. As the rotor speed decreases, more voltage is induced in the rotor windings and more rotor current is developed to create the necessary torque at the reduced speed.

If all of the resistance is removed from the rotor circuit, the current and the motor speed will increase. However, the rotor speed always will be less than the synchronous speed of the field developed by the stator windings. Recall that this fact also is true of the squirrel-cage induction motor. The speed of a wound-rotor motor can be controlled manually or automatically with timing relays, contactors, and pushbutton speed selection.

Torque Performance   As a load is applied to the motor, both the percent slip of the rotor and the torque developed in the rotor increase. As shown in Figure 5–9, the relationship between the torque and percent slip is practically a straight line.

Figure 5–9 illustrates that the torque performance of a wound-rotor induction motor is good whenever the full resistance of the speed controller is inserted in the rotor circuit. The large amount of resistance in the rotor circuit causes the rotor current to be almost in phase with the induced voltage of the rotor. As a result, the field set up by the rotor current is almost in phase with the stator field. If the two fields reach a maximum value at the same instant, there will be a strong magnetic reaction, resulting in a high torque output.

However, if all of the speed controller resistance is removed from the rotor circuit and the motor is started, the torque performance is poor. The rotor circuit minus the speed controller resistance consists largely of inductive reactance. This means that the rotor current lags behind the induced voltage of the rotor and, thus, the rotor current lags behind the stator current. As a result, the rotor field set up by the rotor current lags behind the

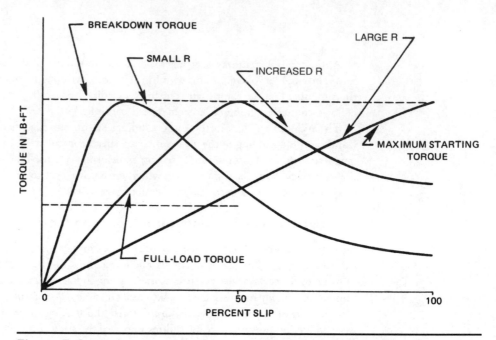

Figure 5–9 Performance curves of a wound-rotor motor; Courtesy Alerich, *Electricity 4,* Delmar Publishers.

stator field, which is set up by the stator current. The resulting magnetic reaction of the two fields is relatively small, since they reach their maximum values at different points. In summary, then, the starting torque output of a wound-rotor induction motor is poor when all resistance is removed from the rotor circuit.

Speed Regulation   It was shown in the previous paragraphs that the insertion of resistance at the speed controller improves the starting torque of a wound-rotor motor at low speeds. However, there is an opposite effect at normal speeds. In other words, the speed regulation of the motor is poorer when resistance is added in the rotor circuit at a higher speed. For this reason, the resistance of the speed controller is removed as the motor comes up to its rated speed.

Figure 5–10 shows the speed performance of a wound-rotor induction motor. Note that the speed characteristic curve resulting when all of the resistance is cut out of the speed controller indicates relatively good speed regulation. The second speed characteristic curve, resulting when all of the resistance is inserted in the speed controller, has a marked drop in speed as the load increases. This indicates poor speed regulation.

Power Factor   The power factor of a wound-rotor induction motor at no load is as low as 15 to 20 percent lag. However, as load is applied to the motor, the power factor improves and increases to 85 to 90 percent lag at rated load.

Figure 5–11 is a graph of the power factor performance of a wound-rotor induction motor from a no-load condition to full load. The low lagging power factor at no load is due to the fact that the magnetizing component of load current is such a large part of the total motor current. The magnetizing component of load current magnetizes the iron, causing interaction between the rotor and the stator, by mutual inductance.

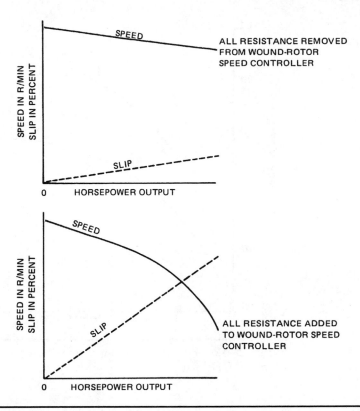

Figure 5–10   Speed performance curves of a wound-rotor motor; Courtesy Alerich, *Electricity 4,* Delmar Publishers.

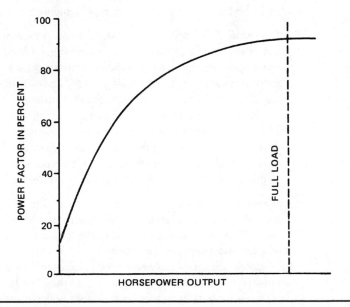

Figure 5–11   Power factor curve of a wound-rotor motor; Courtesy Alerich, *Electricity 4,* Delmar Publishers.

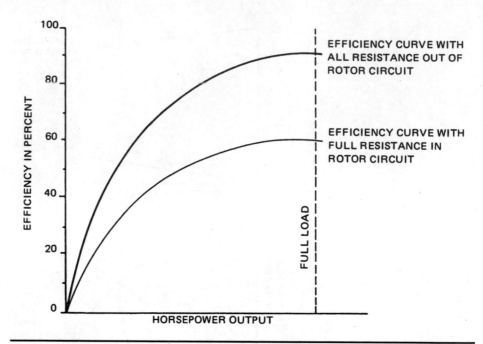

Figure 5–12 Efficiency curve for a wound-rotor motor; Courtesy Alerich, *Electricity 4,* Delmar Publishers.

As the mechanical load on the motor increases, the in-phase component of current increases to supply the increased power demands. The magnetizing component of the current remains the same, however. Since the total motor current is now more nearly in phase with the line voltage, there is an improvement in the power factor.

Operating Efficiency Both a wound-rotor induction motor with all of the resistance cut out of the speed controller and a squirrel-cage induction motor show nearly the same efficiency performance. However, when a motor must operate at slow speeds with all of the resistance cut in the rotor circuit, the efficiency of the motor is poor because of the power loss in watts in the resistors of the speed controller.

Figure 5–12 illustrates the efficiency performance of a wound-rotor induction motor. The upper curve showing the highest operating efficiency results when the speed controller is in the fast position and there is no resistance inserted in the rotor circuit. The lower curve shows a lower operating efficiency. This occurs when the speed controller is in the slow position and all of the controller resistance is inserted in the rotor circuit.

Reversing Rotation The direction of rotation of a wound-rotor induction motor is reversed by interchanging the connections of any two of the three line leads, as in Figure 5–13. This procedure is identical to the procedure used to reverse the direction of rotation of a squirrel-cage induction motor.

An electrician should never attempt to reverse the direction of rotation of a wound-rotor induction motor by interchanging any of the leads feeding from the slip rings to the speed controller. Changes in these connections will not reverse the direction of rotation of the motor.

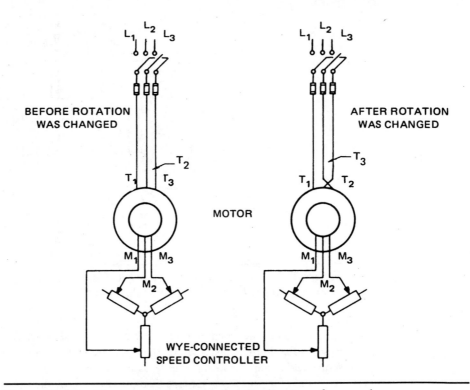

Figure 5–13   Changes necessary to reverse the direction of a wound-rotor motor; Courtesy Alerich, *Electricity 4,* Delmar Publishers.

The Synchronous Motor

The synchronous motor is a three-phase AC motor that operates at a constant speed from a no-load condition to full load. This type of motor has a revolving field that is separately excited from a direct-current source. In this respect, it is similar to a three-phase AC generator. If the DC field excitation is changed, the power factor of a synchronous motor can be varied over a wide range of lagging and leading values.

The synchronous motor is used in many industrial applications because of its fixed speed characteristic over the range from no load to full load. This type of motor also is used to correct or improve the power factor of three-phase AC industrial circuits, thereby reducing operating costs.

Construction Details

A three-phase synchronous motor basically consists of a stator core with a three-phase winding (similar to an induction motor), a revolving DC field with an auxiliary or amortisseur winding, and slip rings, brushes and brush holders, and two end shields housing the bearings that support the rotor shaft. An amortisseur winding consists of copper bars embedded in the cores of the poles. The copper bars of this special type of "squirrel-cage winding" are welded to end rings on each side of the rotor.

Both the stator winding and the core of a synchronous motor are similar to those of the three-phase, squirrel-cage induction motor and the wound-rotor induction motor. The leads for the stator winding are marked T^1, T^2, and T^3, and terminate in an outlet box mounted on the side of the motor frame.

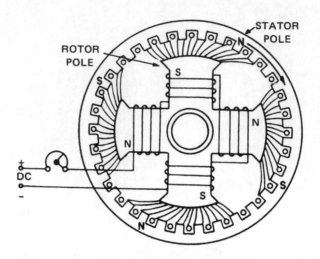

Figure 5–14 Diagram to show principle of operation of a synchronous motor; Courtesy Alerich, *Electricity 4*, Delmar Publishers.

The rotor of the synchronous motor has salient field poles. The field coils are connected in series for alternate polarity. The number of rotor field poles must equal the number of stator field poles. The field circuit leads are brought out to two slip rings mounted on the rotor shaft for brush-type motors. Carbon brushes mounted in brush holders make contact with the two slip rings. The terminals of the field circuit are brought out from the brush holders to a second terminal box mounted on the frame of the motor. The leads for the field circuit are marked F^1 and F^2. A squirrel-cage, or *amortisseur*, winding is provided for starting because the synchronous motor is not self-starting without this feature. The rotor shown in Figure 5–14 has salient poles and an amortisseur winding.

Two end shields are provided on a synchronous motor. One of the end shields is larger than the second shield because it houses the DC brush holder assembly and slip rings. Either sleeve bearings or ball-bearing units are used to support the rotor shaft. The bearings also are housed in the end shields of the motor.

Principle of Operation

When the rated three-phase voltage is applied to the stator windings, a rotating magnetic field is developed. This field travels at the synchronous speed. As stated previously, the synchronous speed of the magnetic field depends on the frequency of the three-phase voltage and the number of stator poles using the formula:

$$S = \frac{120 \times f}{p}$$

The magnetic field, which is developed by the stator windings, travels at synchronous speed and cuts across the squirrel-cage winding of the rotor. Both voltage and current are induced in the bars of the rotor winding. The resulting magnetic field of the amortisseur (squirrel-cage) winding reacts with the stator field to create a torque, which causes the rotor to turn.

The rotation of the rotor will increase in speed to a point slightly below the synchronous speed of the stator, about 92 to 97 percent of the motor-rated speed. There is a small

ATTRACTION BETWEEN UNLIKE POLES

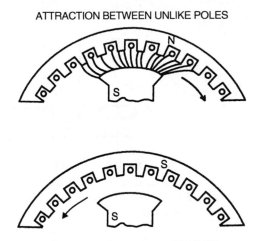

REPULSION BETWEEN LIKE POLES

Figure 5–15   Starting of a synchronous motor; Courtesy Alerich, *Electricity 4,* Delmar Publishers.

slip in the speed of the rotor behind the speed of the magnetic field set up by the stator. In other words, the motor is started as a squirrel-cage induction motor.

The field circuit is now excited from an outside source of direct current, and fixed magnetic poles are set up in the rotor field cores. The magnetic poles of the rotor are attracted to unlike magnetic poles of the magnetic field set up by the stator.

Figures 5–14 and 5–15 show how the rotor field poles lock with unlike poles of the stator field. Once the field poles are locked, the rotor speed becomes the same as the speed of the magnetic field set up by the stator windings. In other words, the speed of the rotor is now equal to the synchronous speed.

Remember that a synchronous motor must always be started as a three-phase, squirrel-cage induction motor with the DC field excitation disconnected. The DC field circuit is added only after the rotor accelerates to a value near the synchronous speed. The motor then will operate as a synchronous motor, locked in step with the stator rotating field.

If an attempt is made to start a three-phase synchronous motor by first energizing the DC field circuit and then applying the three-phase voltage to the stator windings, the motor will not start, since the net torque is zero. At the instant the three-phase voltage is applied to the stator windings, the magnetic field set up by the stator current turns at the synchronous speed. The rotor, with its magnetic poles of fixed polarity, is attracted first by an unlike stator pole and attempts to turn in that direction. However, before the rotor can turn, another stator pole of opposite polarity moves into position and the rotor then attempts to turn in the opposite direction. Because of this action of the alternating poles, the net torque is zero and the motor does not start.

Direct-Current Field Excitation   In the early models, the field circuit was excited from an external direct-current source. A DC generator may be coupled to the motor shaft to supply the DC excitation current.

Figure 5–16 shows the connections for a synchronous motor. A field rheostat in the separately excited field circuit varies the current in the field circuit. Changes in the field current affect the strength of the magnetic field developed by the revolving rotor. Variations

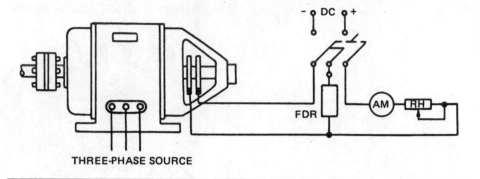

Figure 5–16   External connection for a synchronous motor; Courtesy Alerich, *Electricity 4*, Delmar Publishers.

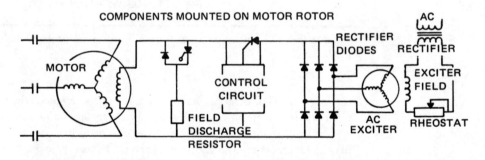

Figure 5–17   Simplified circuit for a brushless synchronous motor; Courtesy Alerich, *Electricity 4*, Delmar Publishers.

in the rotor field strength do not affect the motor, which continues to operate at a constant speed. However, changes in the DC field excitation do change the power factor of a synchronous motor.

Brushless Solid-State Excitation   An improvement in synchronous motor excitation is the development of the brushless DC exciter. The commutator of a conventional direct-connected exciter is replaced with a three-phase, bridge-type, solid-state rectifier. The DC output is then fed directly to the motor field winding. Simplified circuitry is shown in Figure 5–17. Shaded components are mounted on the rotor. A stationary field ring for the AC exciter receives DC from a small rectifier in the motor control cabinet. This rectifier is powered from the AC source. The exciter DC field is also adjustable. Rectifier solid-state diodes change the exciter AC output to DC. This DC is the source of excitation for the rotor field poles. Silicon-controlled rectifiers, activated by the solid-state field control circuit, replace electromechanical relay and the contactors of the conventional brush-type synchronous motor.

The field discharge resistor is inserted during motor starting. At motor synchronizing pull-in speed, the field discharge circuit is automatically opened and DC excitation is applied to the rotor field pole windings. Excitation is automatically removed if the motor pulls out of step (synchronization) due to an overload or a voltage failure. The stator of a brushless motor is similar to that of a brush-type motor. The rotor, however, differs. Mounted on the rotor shaft are the armature of the AC exciter, the AC output of which is rectified to DC by the silicon diodes. Brush and commutator problems are eliminated with this system.

Power Factor   A poor lagging power factor results when the field current is decreased below normal by inserting all of the resistance of the rheostat in the field circuit. The three-phase AC circuit to the stator supplies a magnetizing current, which helps strengthen the weak DC field. This magnetizing component of current lags the voltage by 90 electrical degrees.

Since the magnetizing component of current becomes a large part of the total current input, a low lagging power factor results.

If a weak DC field is strengthened, the power factor improves. As a result, the three-phase AC circuit to the stator supplies less magnetizing current. The magnetizing component of current becomes a smaller part of the total current input to the stator winding, and the power factor increases. If the field strength is increased sufficiently, the power factor increases to unity or 100 percent. When a power factor value of unity is reached, the three-phase AC circuit does not supply any current and the DC field circuit supplies all of the current necessary to maintain a strong rotor field. The value of DC field excitation required to achieve unity power factor is called *normal field excitation*.

If the magnetic field of the rotor is strengthened further by increasing the DC field current above the normal excitation value, the power factor decreases. However, the power factor is leading when the DC field is overexcited. The three-phase AC circuit feeding the stator winding delivers a demagnetizing component of current that opposes the too-strong rotor field. This action results in a weakening of the rotor field to its normal magnetic strength.

Figure 5–18 shows how the DC field is aided or opposed by the magnetic field set up by the AC windings. It is assumed in Figure 5–18 that the DC field is stationary and a revolving armature is connected to the AC source. The student should keep in mind the fact that most synchronous motors have stationary AC windings and a revolving DC field. For either case, however, the principle of operation is the same.

Figure 5–19 shows two characteristic operating curves for a three-phase synchronous motor. With normal full-field excitation, the power factor has a peak value of unity or 100 percent and the AC stator current is at its lowest value. As the DC field current is decreased in value, the power factor decreases in the lag quadrant and there is a resulting rapid rise in the AC stator current. If the DC field current is increased above the normal field excitation value, the power factor decreases in the lead quadrant and a rapid rise in the AC stator current results.

It has been shown that a synchronous motor operated with an overexcited DC field has a leading power factor. For this reason, a three-phase synchronous motor often is connected to a three-phase industrial feeder circuit having a low lagging power factor (Figure 5–20). In other words, the synchronous motor with an overexcited DC field will correct the power factor of the industrial feeder circuit.

In Figure 5–20, two induction motors with lagging power factors are connected to an industrial feeder circuit. The synchronous motor connected to the same feeder is operated with an overexcited DC field. Since the synchronous motor can be adjusted so that the resulting power factor is leading, the power factor of the industrial feeder can be corrected until it reaches a value near unity or 100 percent.

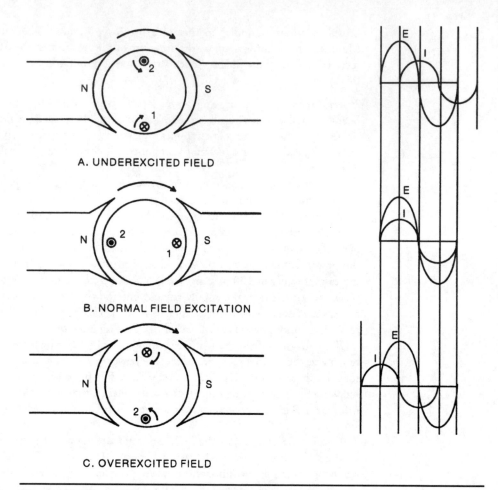

A. UNDEREXCITED FIELD

B. NORMAL FIELD EXCITATION

C. OVEREXCITED FIELD

Figure 5–18 Field excitation in a synchronous motor; Courtesy Alerich, *Electricity 4,* Delmar Publishers.

Reversing Rotation The direction of rotation of a synchronous motor is reversed by interchanging any two of the three line leads feeding the stator winding. The electrician must remember that the direction of rotation of the motor does not change if the two conductors of the DC source are interchanged.

Industrial Applications

The three-phase synchronous motor is used when a prime mover having a constant speed from a no-load condition to full load is required, such as fans, air compressors, and pumps. The synchronous motor is used in some industrial applications to drive a mechanical load and also correct power factor. In some applications, this type of motor is used only to correct the power factor of an industrial power system. When the synchronous motor is used only to correct the power factor and does not drive any mechanical load, it serves the same purpose as a bank of capacitors used for power factor correction. Therefore, in such an installation the motor is called a synchronous capacitor.

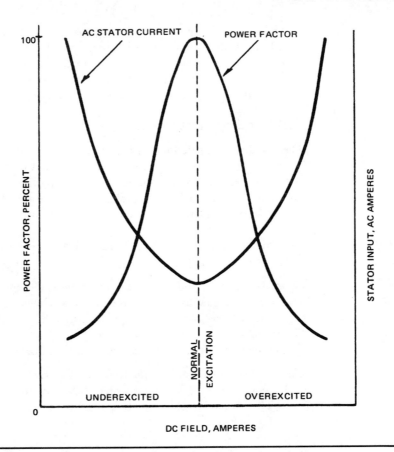

Figure 5–19   Characteristic operating curves for a synchronous motor; Courtesy Alerich, *Electricity 4,* Delmar Publishers.

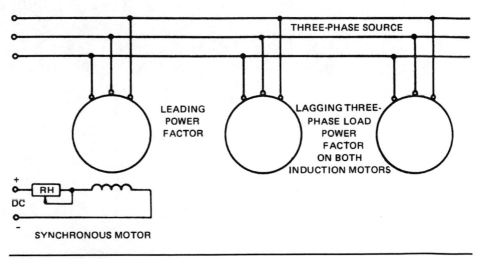

Figure 5–20   Synchronous motors used for power factor correction; Courtesy Alerich, *Electricity 4,* Delmar Publishers.

Three-phase synchronous motors, up to a rating of 10 horsepower, usually are started directly across the rated three-phase voltage. Synchronous motors of larger sizes are started through a starting compensator or an automatic starter. In this type of starting, the voltage applied to the motor terminals at the instant of start is about half the value of the rated line voltage and the starting surge of current is limited.

Part-Winding-Start Motor

The windings of a part winding motor are arranged so that part of the primary winding can be energized and then other parts are energized in stages. This motor is used in the same situation you may use reduced voltage starting. This motor is ideal when starting high torque and current would cause a voltage drop to the distribution system or in plant equipment. The wye-delta starting is associated with the part-winding-start motor.

Identifying the Nine Leads of Untagged Three-Phase Motors

Identifying the Leads of a Three-Phase, Wye-Connected, Dual-Voltage Motor

Figure 5–21 shows standardized terminal markings. These markings can be connected to a 240-volt line. Figure 5–22 shows the motor windings connected parallel for low voltage applications. Series connections apply to high voltage applications such as 480 volts, see Figure 5–23.

Use the following steps of action to identify unmarked leads:

Use an Ohmmeter to Perform the Following Checks

1. In a wye motor, three leads should show continuity T_9, T_7, T_8 in Figure 5–24. Three other sets of leads will show continuity T_3 and T_6, T_4 and T_1, T_5 and T_2 in Figure 5–21.

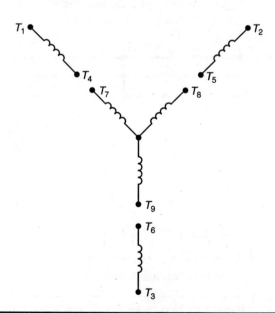

Figure 5–21 Standardized terminal markings.

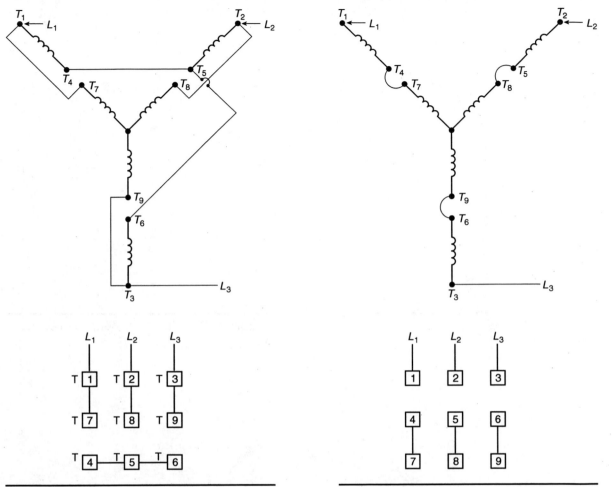

Figure 5–22   Low voltage connection.

Figure 5–23   High voltage connection.

2. Mark the three leads T_7, T_8, and T_9. Connect T_7, T_8 and T_9 to a 240-volt source as shown in Figure 5–25. (*Note:* These windings can be energized without damage occurring to the motor, provided no load is connected.)

3. With the power removed, connect one end of the paired leads, T_3 and T_6, T_4 and T_1, T_5 and T_2, to the T_7 lead. Energize T_9, T_7 and T_8 with 480 volts, as described in Figure 5–25.

Use a Voltmeter to Perform the Following Checks

1. If voltages are unequal, the wrong lead is connected to T_7. Remove power and change lead. When the proper lead is connected, the voltages at leads T_8 and T_9 will be equal.

2. When all paired leads are found, if T_4 is connected to T_7 you should read approximately 360 volts between T_1 and T_8 or T_9 as described in Figure 5–27. If T_1 is connected to T_7, your voltage will be approximately 140 volts between T_4 and T_8 or T_4 and T_9, as described in Figures 5–26 and 5–27. Remember that

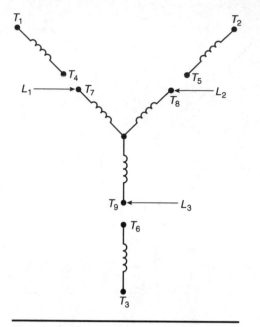

Figure 5–24 Step 2—connect L_1 to T_7; L_2 to T_8; L_3 to T_9.

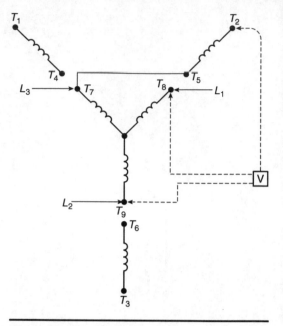

Figure 5–25 Step 3—measure voltage at pairs not connected to T_8 and T_9.

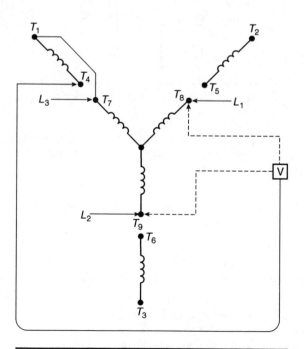

Figure 5–26 Step 4—connect T_1 to T_7.

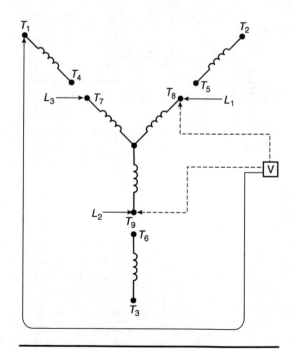

Figure 5–27 Step 5—connect T_4 to T_7.

this is an inductive motor and is additive and subtractive depending on how the leads are connected.

Identifying the Leads of a Three-Phase, Delta-Connected Motors Leads

Delta-connected motors often are used in larger horsepower situations. Some motor starting schemes use a wye-delta (star-delta) starting system (if the motor is designed to run on a delta in a different way than the wye-connected motors). As mentioned earlier, the delta motor has three sets of three leads with continuity, compared to the star motor, which has one set of leads (see Figure 5–28).

Use an Ohmmeter to Perform the Following Checks

1. Find three sets of three leads that have continuity (see Figure 5–29). Temporarily tag the center lead as T_1 and the remaining two leads of the group T_4 and T_9.

2. In the next coil group, tag the center T_2 and the remaining two leads of the group as T_5 and T_7. For the last coil group, tag the center lead T_3 and the remaining two leads of the group as T_6 and T_8.

3. Make sure that all coils are isolated in an open position when energizing the coils. Open leads will be energized when the coils are hot.

Use a Voltmeter to Perform the Following Checks

1. Connect the first coil group T_1, T_4, T_9 to the lowest voltage indicated on the nameplate: L_1 to T_1, L_2 to T_4, L_3 to T_9. Energize the coils and run the motor. Note direction of rotation.

2. Connect the lead marked T_4 to the lead you labeled T_7. Leave the line leads connected, as in step 1. Read the voltage between T_1 and T_2. If the lead markings are correct, voltage T_1 to T_2 should be approximately twice the applied line voltage. If it only reads 1.5 times the low voltage, reconnect T_4 to the lead marked T_8.

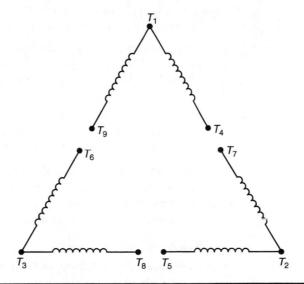

Figure 5–28 Standard terminal markings.

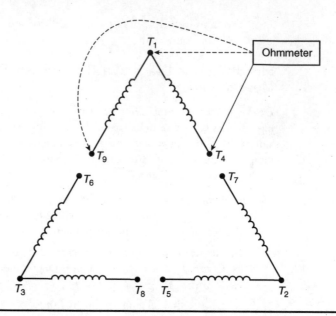

Figure 5–29 Use ohmmeter to check continuity of three leads—T_3 to T_6 to T_8, T_5 to T_2 to T_7, T_1 to T_4 to T_9.

3. Measure T_1 to T_2 again. If the voltage is twice the line voltage, the markings should be changed to indicate that T_4 is now connected to T_7. If the voltage measured T_1 to T_2 is less than line voltage, then reconnect T_9 to T_7, essentially reversing both coil leads. The final connection should read twice the line voltage and T_4 should be connected to T_7.

4. Connect the third coil group to the first coil group. Use the first coil group as the reference point. Connect the coils until the lead connected to T_9 of the first group reads twice the applied voltage from T_1 to T_3. The lead connections that indicate this voltage determine that T_9 is connected to T_6 and the other loose end is T_8.

5. Mark each line lead as L_1, L_2, L_3, to check each of your lead markings. Connect L_1 to T_2, L_2 to T_5, L_3 to T_7. Energize just the second coil group from the line. The motor should run the same direction as originally connected with the first coil group.

6. The last check is to verify the third group. Connect L_1 to T_3, L_2 to T_6, L_3 to T_8. The motor should rotate in the same direction as the original group. If not, recheck your markings before connecting all the coils together.

If all the tests and checks are done and the lead markings are established, connect the motor for both low and high voltage connections, as shown in Figures 5–30 and 5–31. The motor should run smoothly with no noise and the current in all three lines should be equal or less than the nameplate current, if the motor is unloaded.

Trouble-shooting Checklist and Solutions

1. **Problem: Motor will not start.**

   • **Probable cause:** Branch circuit or feeder overcurrent protection device is open (fuse or circuit breaker). Overcurrent protection opens because a ground fault, a short circuit, or both have occurred. Check the circuit for possible faults (broken, damaged wire, or components) before replacing or resetting the device.

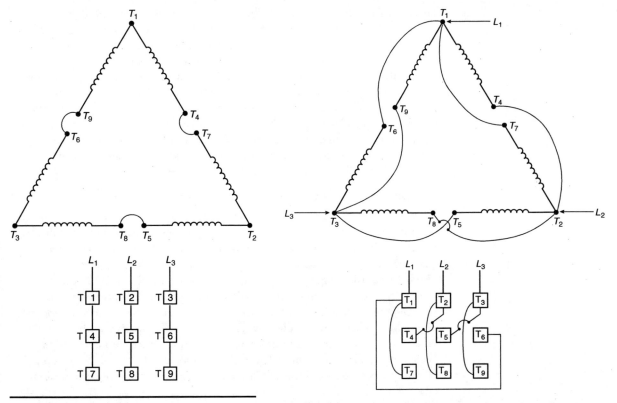

Figure 5–30 Step 7—high voltage connection.

Figure 5–31 Step 8—Low voltage connection.

- **Probable cause:** Tripped overload. Check the overload adjustment if applicable. Perform amp draw test at motor or starter. Replace or reset the overload.
- **Probable cause:** Single phasing condition. One of the phases in the three-phase system has opened. If this event occurs while the motor is running, it will continue to turn. If the event occurred prior to starting, the motor will not start.
- **Probable cause:** Short circuit in winding. This condition occurs when the coil insulation is damaged. This condition will eventually burn into open. With a visual inspection or test equipment this condition can be found. See testing section for proper procedures. If the coil insulation integrity is compromised, rewind the coil.
- **Probable cause:** Ground fault in winding. This occurs when the damaged winding touches the case of the motor or shaft. This problem can be found with a visual inspection or by using test equipment. See Chapter 9 for testing procedure. Repair by rewinding the coil.
- **Probable cause:** Broken or damaged rotor bars. This condition will cause the motor to lose its ability to perform under a load. Check the motor with an amp draw test under a no-load condition. A low ampere rating would indicate a damaged rotor. Add a light load to the motor; if the motor speed decreases as load is added, this would indicate a damaged rotor. Check the motor at full load; if the voltage is below the nameplate rating (check the voltage tolerance level, normally not less than 5 percent), the rotor is probably damaged. Hairline

fractures are difficult to see with the naked eye. You will need special test equipment (growler) to locate this problem. See Chapter 9 for testing procedure.

- **Probable cause:** Damaged controller. Damaged wire, contacts, coils, and connections can cause the controller not to operate properly.
- **Probable cause:** Loose, worn, frozen, or bad bearings. Bearings keep the rotor from touching the stator winding. If the rotor touches the stator winding, a short circuit will occur due to the wear of the turning action. Test the bearings by inserting a feeler gauge in the air gaps. Air gaps should be equal at all points. If the air gaps are uneven, bearings are worn and should be replaced. When replacing the bearings, check the armature and stator windings for worn or bare places. Repair all damaged surfaces of the armature and stator.

2. **Problem: Motor starts but does not run properly.**
- **Probable cause:** Branch or feeder overcurrent protection device is open. Fuse or circuit breaker. Overcurrent protection opens because either a ground fault, short circuit, or both has occurred. Check the circuit for possible faults (broken, damaged wire, or components) before replacing or resetting the device.
- **Probable cause:** Single phasing condition. One of the phases in the three-phase system has opened. If this event occurs while the motor is running, it will continue to turn. If the event occurred prior to starting, the motor will not start.
- **Probable cause:** Short circuit in winding. This condition occurs when the coil insulation is damaged. This condition will eventually burn into open. With a visual inspection or test equipment, this condition can be found. See Chapter 9 for proper procedures. If the coil insulation integrity is compromised, rewind the coil.
- **Probable cause:** Ground fault in winding. This occurs when the damaged winding touches the case of the motor or shaft. This problem can be found with a visual inspection or by using test equipment. See Chapter 9 for testing procedure. Repair by rewinding the coil.
- **Probable cause:** Broken or damaged rotor bars. This condition will cause the motor to lose its ability to perform under a load. Check the motor with an amp draw test under a no-load condition. A low ampere rating would indicate a damaged rotor. Add a light load to the motor. If the motor speed decreases as load is added, this would indicate a damaged rotor. Check the motor at full load. If the voltage is below the nameplate rating (check the voltage tolerance level, normally not less than 5 percent), the rotor is probably damaged. Hairline fractures are difficult to see with the naked eye. You will need special test equipment (growler) to locate this problem. See Chapter 9 for testing procedure.
- **Probable cause:** Too low/high voltage. Every motor has a certain amount of low/high voltage it can handle without damage. Motors with several marked voltage ratings on the nameplate can usually handle too low/high voltages. See manufacturer specifications for exact tolerance levels. A general rule of thumb is no less than ±5 percent.
- **Probable cause:** Loose, worn, frozen, or bad bearings. Bearings keep the rotor from touching the stator winding. If the rotor touches the stator windings, a short circuit will occur due to the wear of the turning action. Test the bearings by inserting a feeler gauge in the air gaps. Air gaps should be equal at all points.

If the air gaps are uneven, bearings are worn and should be replaced. When replacing the bearings, check the armature and stator windings for worn or bare places. Repair all damaged surfaces of the armature and stator.

3. **Problem: Motor starts and accelerates slowly.**

   - **Probable cause:** Short circuit in winding. This condition occurs when the coil insulation is damaged. This condition will eventually burn into open. With a visual inspection or test equipment, this condition can be found. See Chapter 9 for proper procedures. If the coil insulation integrity is compromised, rewind the coil.

   - **Probable cause:** Broken or damaged rotor bars. This condition will cause the motor to lose its ability to perform under a load. Check the motor with an amp draw test under a no-load condition. A low ampere rating would indicate a damaged rotor. Add a light load to the motor. If the motor speed decreases as load is added, this would indicate a damaged rotor. Check the motor at full load. If the voltage is below the nameplate rating (check the voltage tolerance level, normally not less than 5 percent), the rotor is probably damaged. Hairline fractures are difficult to see with the naked eye. You will need special test equipment (growler) to locate this problem. See Chapter 9 for testing procedure.

   - **Probable cause:** Tripped overload. Check the overload adjustment if applicable. Perform amp draw test at motor or starter. Replace or reset the overload.

   - **Probable cause:** Loose, worn, frozen, or bad bearings. Bearings keep the rotor from touching the stator winding. If the rotor touches the stator windings, a short circuit will occur due to the wear of the turning action. Test the bearings by inserting a feeler gauge in the air gaps. Air gaps should be equal at all points. If the air gaps are uneven, bearings are worn and should be replaced. When replacing the bearings, check the armature and stator windings for worn or bare places. Repair all damaged surfaces of the armature and stator.

4. **Problem: Motor runs hot.**

   - **Probable cause:** Tripped overload. Check the overload adjustment if applicable. Perform amp draw test at motor or starter. Replace or reset the overload.

   - **Probable cause:** Short circuit in winding. This condition occurs when the coil insulation is damaged. This condition will eventually burn into open. With a visual inspection or test equipment, this condition can be found. See testing section for proper procedures. If the coil insulation integrity is compromised, rewind the coil.

   - **Probable cause:** Broken or damaged rotor bars. This condition will cause the motor to lose its ability to perform under a load. Check the motor with an amp draw test under a no-load condition. A low ampere rating would indicate a damaged rotor. Add a light load to the motor. If the motor speed decreases as load is added, this would indicate a damaged rotor. Check the motor at full load. If the voltage is below the nameplate rating (check the voltage tolerance level, normally not less than 5 percent), the rotor is probably damaged. Hairline fractures are difficult to see with the naked eye. You will need special test equipment (growler) to locate this problem. See Chapter 9 for testing procedure.

- **Probable cause:** Single phasing condition. One of the phases in the three-phase system has opened. If this event occurs while the motor is running, it will continue to turn. If the event occurred prior to starting, the motor will not start.
- **Probable cause:** Loose, worn, frozen, or bad bearings. Bearings keep the rotor from touching the stator winding. If the rotor touches the stator windings, a short circuit will occur due to the wear of the turning action. Test the bearings by inserting a feeler gauge in the air gaps. Air gaps should be equal at all points. If the air gaps are uneven, bearings are worn and should be replaced. When replacing the bearings, check the armature and stator windings for worn or bare places. Repair all damaged surfaces of the armature and stator.

CHAPTER 6

Direct-Current Motors*

Construction

The DC motor consists of the following parts, as shown in Figure 6–1:

Armature—Armatures are the rotating parts of the motor. Armature coils are wound several ways. The way the motor is wound is based on the intended use of the motor.

Commutator—he commutator rides on the shaft of the armature. The commutator serves to convey voltage induced by the brushes. The commutator consists of segments of copper separated by mica insulation. The mica is lower than the copper segments to prevent a short circuit. Slots are cut into the commutator so that solder connections can connect the commutator to the armature coils.

Brushes—Brushes are pressed against the commutator and held by a spring located in the brush holder. Brushes are isolated from the housing to prevent short circuits and ground faults. Brushes are made of a graphite-type substance. Brushes tend to wear out quicker than most other parts of the motor. This wear leaves a residue on the commutator that insulates the commutator from the brushes, thus stopping current flow.

End plates—End plates and bell housing are used interchangeably; they bear the weight of the armature and house the bearings.

Field poles—The field poles are attached to the frame of the motor and are used as a means of support for the field windings. The field windings are isolated from the pole by tape and insulating wire.

Frame—The frame of the motor attaches the bell housing of the motor and usually is made of cast iron.

Principles of DC Motor Operation

As the armature rotates, the magnetic field produced by the field windings is cut by the magnetic field produced by the armature windings, thus generating a voltage. This voltage is opposite to the source (sometimes called *applied* or *line*) voltage. This voltage is called

*Written with permission from Alerich, *Electricity 4*, Delmar Publishers.

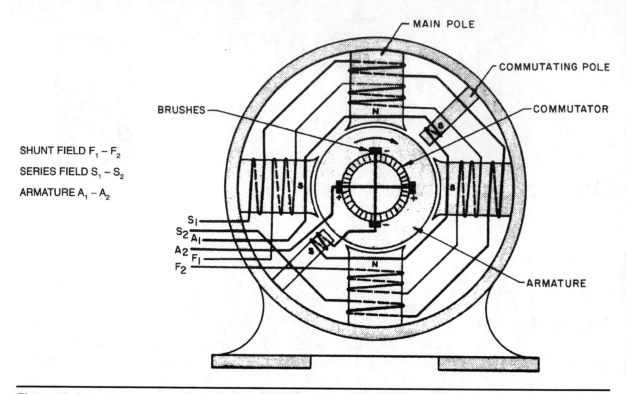

SHUNT FIELD $F_1 - F_2$

SERIES FIELD $S_1 - S_2$

ARMATURE $A_1 - A_2$

Figure 6–1 DC generator with four poles and four brushes makes the DC output smoother; Courtesy Alerich, *Electricity 4,* Delmar Publishers.

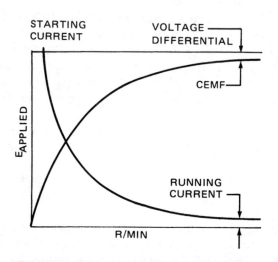

Figure 6–2 Voltage differential is small at no load, creating a small effective voltage; Courtesy Alerich, *Electricity 3,* Delmar Publishers.

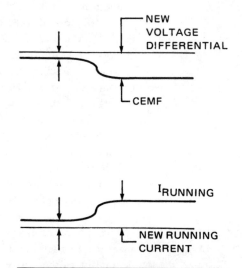

Figure 6–3 As load is added, the c.e.m.f. drops and creates a higher voltage differential; Courtesy Alerich, *Electricity 3,* Delmar Publishers.

counterelectromotive force (c.e.m.f.) or *back electromotive force* (back e.m.f.). The characteristics of c.e.m.f. under load and unload conditions are found in Figures 6–2 and 6–3.

The ability of the DC motor to utilize the counter electromotive force makes it a desirable motor with applications that demand constant speed. DC motors are used for several applications because of the torque characteristics, speed, and size.

DC Motor Types

There are basically five types of DC motors:

1. series
2. shunt
3. compound
4. permanent magnet
5. brushless

Series Motors

Series motors use only a series field in series with the armature. Figure 6–4 illustrates how the series field (represented by three loops) is connected in the motor circuit. The mechanical load has a large effect on series motor operations. Because the motor current flows through the armature and the series field, the motor current has a doubled effect on the torque. Referring to Figure 6–5, notice that the series motor has the most torque per amp of any of the DC motors. These motors are often used as traction motors. Motors that will produce a great deal of torque at low speed and decreased torque at higher speed often are used for drive wheels on locomotives and forklifts.

Speed variations are extreme for the series motor. Figure 6–6 shows how the speed curves compare to the other DC motors. At light mechanical load, the speed of the motor is high. In fact, at no load the speed may be high enough to run away or tear the rotor apart by centrifugal force. Series motors should only be run with a load attached. As mechanical load is added, the c.e.m.f. of the rotor decreases and the motor current increases. The torque is produced by the reaction of the series field flux and the armature flux. Because both magnetic fields fluctuate together, there is a wide range of speed and torque that are inversely proportional.

To reverse the series motor, reverse the relationship between the fields. The preferred method is reversing the armature connection. Speed control is accomplished by adjusting the voltage to the motor, both series and armature.

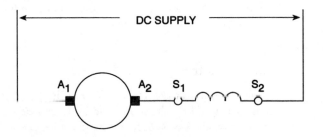

Figure 6–4   Series motor connection diagram; Courtesy Keljik, *Electric Motors and Motor Controls,* Delmar Publishers.

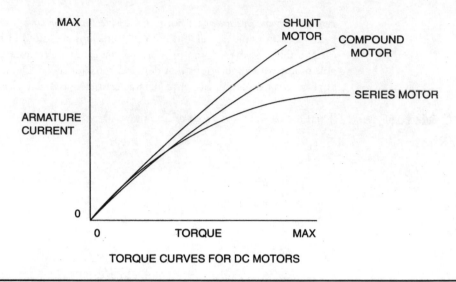

TORQUE CURVES FOR DC MOTORS

Figure 6–5 Comparison of torque versus armature current in DC motors; Courtesy Keljik, *Electric Motors and Motor Controls,* Delmar Publishers.

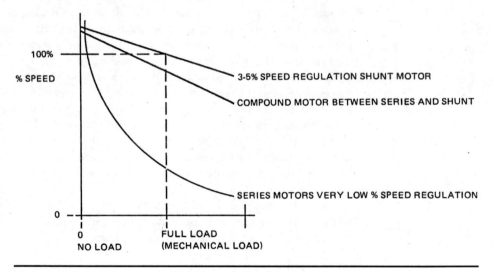

Figure 6–6 Comparison of speed regulation for DC motors; Courtesy Alerich, *Electricity 3,* Delmar Publishers.

Shunt Motors

The shunt motor connection diagram is shown in Figure 6–7. The armature leads are marked A_1 and A_2; the shunt field leads are marked F_1 and F_2. The shunt field is the same type of winding as is found in the shunt generator. It has a relatively high resistance and is designed to be connected across the DC power source. If speed control is needed, a field rheostat may be included. The field rheostat is not used in electronic DC motor control, but the concept of controlling field and armature current is employed.

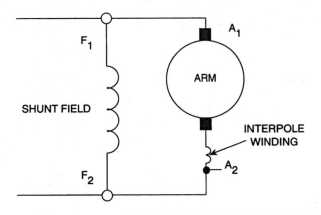

Figure 6–7 Lead marking and connection for a DC shunt motor; Courtesy Keljik, *Electric Motors and Motor Controls,* Delmar Publishers.

The speed of a shunt motor is quite stable; thus, speed regulation is good. It has a low percent speed regulation, so speed stays nearly constant between no- and full-load speed. This is because the shunt field is connected directly to the power source and the magnetic field does not change much. See Figure 6–6 for a comparison graph of speed regulation. If the stator field does not change, the speed and torque are controlled by the armature winding. The lower the armature resistance, the lower the percent speed regulation. Thus, any small change in voltage differential caused by a slowing rotor, due to a heavier mechanical load, causes an increase in developed torque to keep speed constant or nearly constant. Remember, no nonsynchronous motor has 0 percent speed regulation. There must be a change in speed to produce more torque. The torque curves of a shunt motor are almost linear. See Figure 6–5 for torque curve comparisons.

As mentioned, when the motor starts, there is a large DC inrush current flow. This starting current produces enough starting torque to get the rotor spinning. Normally, the inrush current is high enough that DC motor starters are designed to limit this inrush current to about 150 percent. This corresponds to the overcurrent protection table in NEC's article 450–152. For all but instantaneous trip breakers, the percentage used for protection is 150 percent of full-load current.

Speed Control and Direction Shunt motor speed control is accomplished by changing the applied armature voltage. To reduce the normal speed, reduce the voltage. This will reduce the current to the armature leads A_1 and A_2. To increase the speed above nameplate speed, weaken, but do not remove, the shunt field.

Direction of rotation of shunt motors can be changed by reversing the relative direction of the rotor flux, by reversing the armature leads, or by reversing the shunt field leads.

Armature Reaction Armature reaction occurs in DC motors exactly as it does in DC generators. If you recall, armature reaction is the distortion to the original plane of the stator magnetic field. The brushes of the DC motor could be moved in the opposite direction from the direction of rotation each time mechanical load is added and armature current increases. Moving the brushes realigns them in the neutral plane. Brushes for generators are moved in the same direction as the rotor rotation. As in the generator, other windings can be added to the motor's stator to compensate for the distorted field. Commutating poles

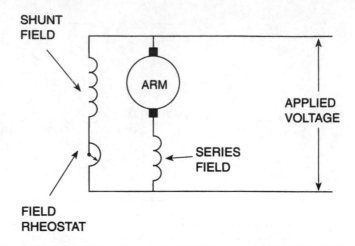

SHUNT
FIELD

ARM

APPLIED
VOLTAGE

SERIES
FIELD

FIELD
RHEOSTAT

Figure 6–8   Schematic of long shunt compound motor; Courtesy Keljik, *Electric Motors and Motor Controls,* Delmar Publishers.

or interpoles are added midway between the salient main poles to shift the magnetic field back to the original neutral plane. Interpoles have the same polarity as the main pole behind them in the direction of rotation. Usually, the interpole windings are connected internally directly in line with the armature, and the leads are brought out at A_1 or A_2 leads. If the leads are brought out for external connections, they are labeled C_1 and C_2. They might also be labeled S_3 and S_4. Figure 6–7 shows a shunt motor with interpoles. The shunt field is represented by at least four loops in the schematic diagram.

Compound Motors

Compound motors are DC motors that use both series field and shunt field windings (see Figure 6–8). These motors are a compromise between the characteristics of the series and the shunt motors. As with the DC generators, the motor fields can be connected *short shunt,* where the shunt field is shorted across the armature, or *long shunt,* where the shunt field is connected across the armature and the series field. The long shunt connection is the more common motor connection because it has better speed regulation. The two fields can also be connected to aid each other in the cumulative compounded motors or connected to oppose each other in the differentially compounded motor. Cumulative compound, long shunt is the standard connection.

If a motor is mistakenly connected differentially, the shunt field will determine the direction at no load. As load is added and more current flows in the series field, the opposite series field may become stronger and suddenly reverse the rotation. To test the proper direction of each field, connect each field separately and note direction of rotation produced by each. Remember to have a mechanical load connected to the motor when testing the series field connection to prevent a runaway condition. Then reconnect the fields into the cumulative pattern.

To reverse the direction of rotation, change just the armature connections. Figure 6–6 shows how the speed/load curves compare to other DC motors and Figure 6–5 shows how the compound motor compares to other DC motors in torque/current curves.

Permanent Magnet Motors

Permanent magnet (PM) motors are a variation of the shunt motor. Instead of wound fields in the stator, ceramic magnets (permanent magnets) have been installed. Now no power is consumed by the field and the efficiency of the motor increases. In addition, the size of the motor can be reduced. Another advantage is that the PM motor is not susceptible to armature reaction. The permanent magnetic field is not distorted by the armature current. The rotor is essentially the same as the standard shunt motor. The speed is easily controlled by adjusting the applied voltage to the armature through the brushes. Overheating can result if PM motors are operated at continuous high torque levels.

The permanent magnet field means there is less copper loss in the motor; therefore, it is more efficient and runs cooler. This feature also enables the PM motor to be used where portable battery-operated motors are required. Additionally, the permanent magnet system provides some braking of the motor. Dynamic braking can be achieved by shorting the only two power leads (armature leads) after disconnecting power.

Reversing the motor is done by reversing the DC polarities to the armature leads. Permanent magnet motors lose some magnetic field strength at very low temperatures. *Caution:* Because PM motors have high starting torque (175 percent or more), the high torque could damage the gear motor assembly.

Brushless DC Motors

The stator field of a brushless DC motor actually rotates. Commutation of the DC supply voltage is controlled electronically by switching the magnetic polarity of the stator poles to correspond to the position of the permanent magnet rotor. Feedback circuits are needed to determine the actual rotor position. The stator has windings that are energized by a digital sequencer. As the rotor north pole position is tracked, the stator south pole is energized just ahead of the rotor pole, so that the rotor north follows the stator south pole. Speed of the rotation is determined by the pulsing or energizing frequency of the stator fields.

There are several advantages of brushless DC motors.

1. They do not have brushes or commutators to wear out.
2. There is no brush sparking or radio frequency interference (RFI) that accompanies brush sparking.
3. The speed torque curves are linear and there are no armature losses.
4. Higher horsepower can be obtained from physically smaller motors.

CHAPTER 7

Shaded Pole Motors, Universal, Stepper, Hybrid, and Torque Motors*

Shaded Pole Motors

Shaded pole motors are one of the simplest and cheapest motors to construct. The principle of operation uses the effects of induction not only into the squirrel cage rotor, but also into parts of the stator that create a rotating magnetic field from a single phase of input voltage. These motors are typically fractional HP ratings and are used in applications that do not require a great deal of starting torque.

In a simple unidirectional motor there is a ring (shading ring) of solid conductor short-circuited and embedded in one side of a stator winding (see Figure 7–1). As the voltage is applied to the top and bottom coil, the shading coil has voltage induced into it. Lenz's law states that the effect of induction always opposes the cause of induction. Therefore, the magnetic field developed by the shading ring as current flows through its shorted winding opposes the main flux. This causes the main magnetic field to be shifted away from the shading ring.

As the applied voltage waveform begins to decrease from its peak value, the magnetic lines of force also decrease. The effect on the shading ring is the opposite. As main current flow decreases, the magnetic effect of the shading ring tends to keep the same polarity as the main pole, but increases in strength. This causes the stator pole to move from the main (unshaded) pole toward the shaded pole (see Figure 7–2).

Position 1—Shading coil produces south pole, main pole north. (Use Figure 7–1 as reference.)
Position 2—Zero shading coil current, main pole north
Position 3—Shading coil current and main pole both north
Position 4—Main coil has zero flux, shading remains north

This produces a clockwise rotation of north pole flux.

To reverse the direction of rotation on shaded pole motors, the relationship between the shading ring and the pole face must be reversed, as the field always moves from the unshaded to the shaded portion of the stator pole. Another set of shading rings on the opposite side may be shorted.

*Adapted with permission from Keljik, *Electric Motors and Motor Controls*, Delmar Publishers.

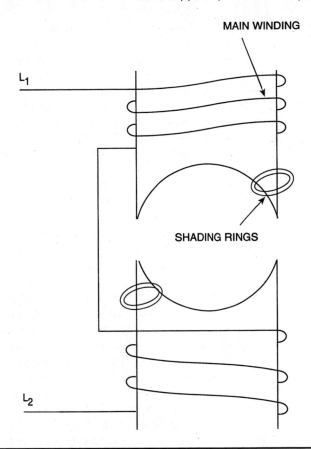

Figure 7–1 Schematic of shaded pole motor; Courtesy Keljik, *Electric Motors and Motor Controls*, Delmar Publishers.

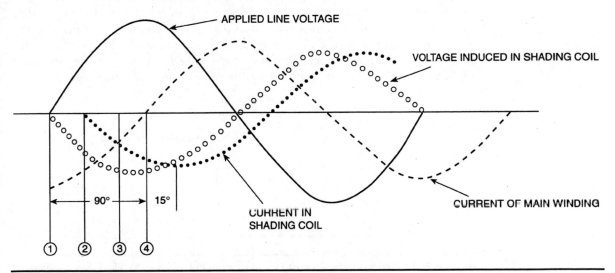

Figure 7–2 Shaded pole motor voltage and current waveforms: line voltage, main coil current, induced shading coil waveforms; Courtesy Keljik, *Electric Motors and Motor Controls*, Delmar Publishers.

Controlling the speed of shaded pole motors is easy. Either alter the voltage that is applied to the stator winding to produce less volts per turn on the winding or change the number of turns by using tapped windings and maintaining the same applied voltage. Either method changes the volts per turn on the motor. Less volts per turn means less flux, more slip, and a slower speed under load.

Shaded pole motors are typically used for fans where the blades are directly mounted to the rotor shaft and the air passes over the motor. These fans require little starting torque, and the air over the motor helps keep it cool. If these motors run hot, or are in a high vibration area, the shading rings (which are soldered rings of conductors) can open up and cause the motor to fail.

Universal Motors

Universal motors are conduction motors. In other words, the current is connected to the rotating rotor through conductors, rather than relying on induction to deliver power to the rotor. The name *universal* indicates that the motor will operate on AC or DC or many frequencies of AC. The operating characteristics are not identical for all waveforms. These motors are sometimes referred to as series motors.

The principle of operation for universal motors is similar to a DC series motor. The rotor is wound with coils (not a squirrel cage), but this is not a wound-rotor motor. (The name *wound-rotor motor* is actually a special three-phase motor, discussed in chapter 5.) Because current is conducted through the rotor windings via a commutator and brushes, the poles can be controlled easily (see Figure 7–3).

A universal motor is designed with laminated iron material to reduce the effects of eddy currents. The iron also is formulated to reduce the effects of hysteresis when alternating current is applied to the coils. Some AC motors use a compensating winding placed 90 degrees electrically from the main winding coils. The compensating winding creates a field that fluctuates with load current to reduce the voltage drop in the armature due to reactance. This helps keep the rotor voltage more constant under varying loads and also reduces the sparking.

To reverse the direction of rotation for the series or universal motor, the relative poles of the rotor and stator must be reversed. To do this, the connections to the brushes at the commutator must be reversed to allow current to flow the opposite way through the rotor while current flows in the same direction in the stator coils. Shift the brushes, after reversing connections to readjust the neutral electrical plane, to reduce sparking. The neutral plane is the position where the armature voltages balance out and create a location where there is the smallest difference in potential and, therefore, the least amount of brush sparking.

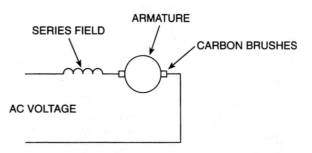

Figure 7–3   Series AC (universal) motor schematic; Courtesy Keljik, *Electric Motors and Motor Controls,* Delmar Publishers.

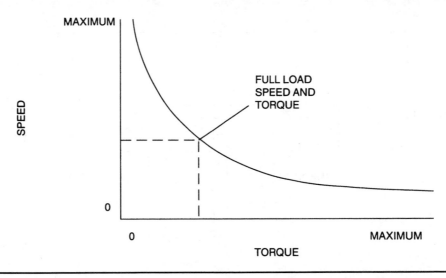

Figure 7–4 Universal motor—speed versus torque curve; Courtesy Keljik, *Electric Motors and Motor Controls,* Delmar Publishers.

Series motors are varying speed motors. This means they have drastic speed changes with load. As the motor is running at no load (no mechanical load), it has very high RPM. The RPM may be so high that the centrifugal force on the rotor begins to pull the commutator and windings off the rotor. Thus, the motor usually is connected to some mechanical load at all times. The load may consist of gear boxes or other devices permanently connected so there is load at all times (for example, drills, saws, vacuum cleaners, and mixers). At full load the speed is very low, but the torque is high (see Figure 7–4). At high load, there is heavy current draw from the line. The breakdown torque of a series motor is approximately 175 percent. The starting torque is extremely high at 300 to 450 percent of running torque. The series motor also has a good operating power factor.

Speed control is common in universal motors. It is accomplished by controlling the voltage applied to the motor. Placing resistors in the circuit or electronic speed control systems that change the applied voltage value. Less applied voltage means less current and less torque developed by the weaker magnetic fields, thus reduced rotational force slows the motor.

Stepper Motors

Stepper motors are specialized motors that also create incremented steps of motion rather than a smooth, unbroken rotation. The basic stepper concept is explained using a permanent magnet on the rotor with two sets of poles as shown in Figure 7–5. As the stator is energized with pulses of DC, the permanent magnet rotor will be repelled or attracted to line up with the stator magnetic poles. The pulses are provided by a stepper controller.

The electronic controller provides timing and sequencing of the motor, but it operates electronically to provide circuit closure as shown in Figure 7 5. For example, by moving switch 1 and 2 to position *A* or *B*, the top poles or the side poles can be reversed. By following the first switch sequence where both switches are set to *A*, the top and right poles become north magnetic polarity and the rotor will align between the poles. Step two changes switch 1 to *B*. If power is left on the top poles, the rotor will align top to bottom. When switch 1 connects to point *B*, the rotor will again move; the south rotor pole

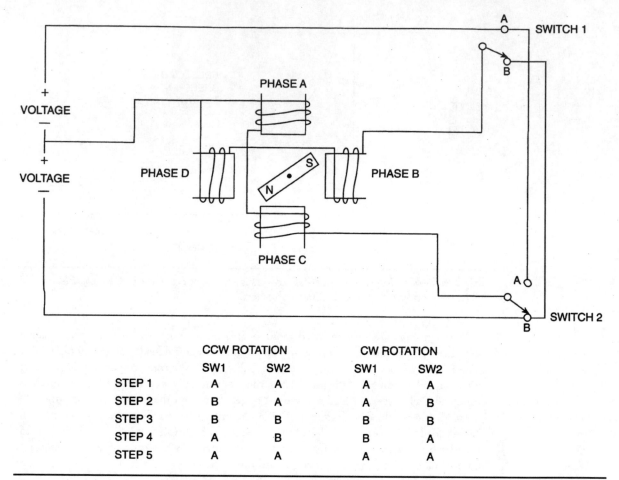

	CCW ROTATION		CW ROTATION	
	SW1	SW2	SW1	SW2
STEP 1	A	A	A	A
STEP 2	B	A	A	B
STEP 3	B	B	B	B
STEP 4	A	B	B	A
STEP 5	A	A	A	A

Figure 7–5   Stepper motor showing switch position and DC voltage source.

aligning between the top and left poles. This will result in CCW rotation. To reverse the direction of rotation, use the second set of steps. Notice that by changing the sequence and length of time the coils are energized, the direction and speed of the steps are controlled.

The rotor could be one of three different styles: the variable reluctance rotor, the permanent magnet rotor, or a combination of the two (a hybrid rotor). Rather than being simply two magnetic poles, the rotor is many magnetic poles, lined up with the teeth on the rotor. The rotor's teeth are spaced so that only one set of teeth remains in perfect alignment with the stator poles at any one time (see Figure 7–6).

By taking the number of times the power must be applied to the stator poles to move one tooth through 360 degrees of mechanical rotation, you can compute the step angle. For example, if the stator needs 200 pulses of power to move one tooth 360 degrees, then divide 360 by 200 to get 1.8 degrees of motion per step. This is the step angle of the motor. The motor will move 1.8 degrees per pulse of stator power. Steppers are available in 90, 45, 15, 7.5, 1.8, and .9 degrees of step angle. The resolution of the motor is a measure of how fine the steps are divided. Resolution is determined by dividing 360 degrees by the step

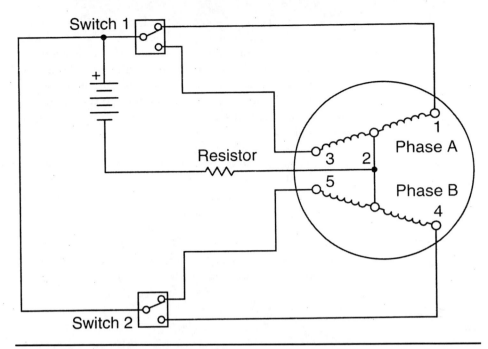

Figure 7–6 Bifilar winding on a stepper motor; Courtesy of Herman, *Standard Textbook of Electricity,* Delmar Publishers.

angle. The 1.8-degree step angle motor takes 200 steps to move around a full revolution, so the resolution is 200. Step angle and resolution are inversely proportional to each other.

The permanent magnet rotor was used for the example on step motion control. These permanent magnet rotors are used with the four pole pieces to provide either 90 or 45 degrees of step angle. This allows the motors to turn at higher speed, but with less resolution. The poles can be physically larger because there are only four of them; thus, the stator windings can carry higher current. Higher current capability means that more torque is available.

Variable Reluctance Rotor

The variable reluctance rotor has the same number of teeth as there are stator poles. The rotor bars (or teeth) are made of high permeability iron. The iron will be magnetized and align with the stator field in small steps. When the rotor is in operation, the rotor retains the magnetism for a short time and will continue to follow the stepped stator field. When the motor is deenergized, the variable reluctance rotor loses its magnetism and does not lock into place as does the permanent magnet rotor.

Hybrid Motors

The hybrid motor uses some of the characteristics of the variable reluctance rotor and some of the permanent magnet rotor characteristics. The stator is wound with two sets of windings on each pole. This method is called the bifilar method. The windings are smaller gauge and have a center tap point brought out for connection. Figure 7–6 shows how a bifilar winding is controlled.

The advantage of this motor is that it provides a very small step angle or high resolution with fewer stator windings. Also, because of the hybrid rotor, it has both strong operating and holding torque for stopped motion. Acceleration and deceleration torques are good because they allow this motor to handle heavier loads at higher stepping speeds, but still maintain accuracy and precision motion control. Most stepping motors are used for precision and incremental motion control. Usually the signals are provided by microprocessor circuitry and are fed to a controller board that processes the signal and applies power to the motor.

Torque Motors

Torque motors are special motors that will deliver maximum torque even if stalled in a locked rotor position. This is useful in applications such as spooling machines or tape drives. They will provide constant tension even if stalled for long periods.

The torque motor can be used where no rotation is required, but only tension applied (similar to a spring). They also can be used to open or close valves that require only a few turns of the motor shaft to be held in position. Torque motors can be used for normal or very slow-speed operations. AC torque motors are normally a polyphase or permanent split capacitor PSC motor design.

CHAPTER

8

Synchronous Motors, Synchros, and Three-Phase Wound-Rotor Induction Motors*

Synchronous Motors

Synchronous motors are three-phase AC motors that operate best at a constant speed from a no-load condition to full load. This type of motor has a rotor that is excited from a direct-current source. The rotor is synchronized (in sync or step) with the rotating magnetic field of the stator windings. A synchronous motor has zero slip, which is contrary to how most induction motors operate. A synchronous motor is similar to a three-phase AC generator. When the DC field excitation is changed, the power factor of a synchronous motor can be varied over a wide range of lagging and leading values. In many cases the synchronous motor helps reduce power factor to an acceptable level.

There are two methods to excite the synchronous motor. The first method is to excite the rotor with direct current. This is usually accomplished with a brushless exciter or DC generator attached to the shaft of the rotor. The stator windings are energized with three-phase power; the motor will run to just below synchronized when the DC excitation occurs. The rotor is energized with DC voltage and increases the speed of the rotor until the magnetic fields of the stator are in sync or step with the rotor. The second method has a rotor that needs no excitation.

To use the synchronous motor for power factor correction of a lagging power factor, you simply overexcite the field windings and cause the motor to draw a larger leading power factor. This leading current compensates for other lagging inductive loads.

The synchronous motor is used in many industrial applications because of its fixed speed characteristic over the range from no load to full load. This type of motor also is used to correct or improve the power factor of three-phase AC industrial circuits, thereby reducing operating costs.

Construction Details

A three-phase synchronous motor basically consists of a stator core with a three-phase winding (similar to an induction motor), a revolving DC field with an auxiliary or amortisseur winding and slip rings, brushes and brush holders, and two end shields housing the

*Adapted with permission from Alerich, *Electricity 4,* Delmar Publishers.

bearings that support the rotor shaft. An amortisseur winding consists of copper bars embedded in the cores of the poles. The copper bars of this special type of "squirrel-cage winding" are welded to end rings on each side of the rotor.

Both the stator winding and the core of a synchronous motor are similar to those of the three-phase, squirrel-cage induction motor and the wound-rotor induction motor. The leads for the stator winding are marked T_1, T_2, and T_3. They terminate in an outlet box mounted on the side of the motor frame.

The rotor of the synchronous motor has salient field poles. The field coils are connected in series for alternate polarity. The number of rotor field poles must equal the number of stator field poles. The field circuit leads are brought out to two slip rings mounted on the rotor shaft for brush-type motors. Carbon brushes mounted in brush holders make contact with the two slip rings. The terminals of the field circuit are brought out from the brush holders to a second terminal box mounted on the frame of the motor. The leads for the field circuit are marked F_1 and F_2. A squirrel-cage, or amortisseur, winding is provided for starting because the synchronous motor is not self-starting without this feature. The rotor shown has salient poles and an amortisseur winding.

Two end shields are provided on a synchronous motor. One end shield is larger because it houses the DC brush holder assembly and slip rings. Either sleeve bearings or ball-bearing units are used to support the rotor shaft. The bearings also are housed in the end shields of the motor.

Principle of Operation

When the rated three-phase voltage is applied to the stator windings, a rotating magnetic field is developed. This field travels at the synchronous speed. As stated in previous units, the synchronous speed of the magnetic field depends on the frequency of the three-phase voltage and the number of stator poles. The following formula is used to determine the synchronous speed.

$$\text{Synchronous speed} = \frac{120 \times \text{frequency}}{\text{number of poles}}$$

$$S = \frac{120 \times f}{P}$$

The magnetic field developed by the stator windings travels at synchronous speed and cuts across the squirrel-cage winding of the rotor. Both voltage and current are induced in the bars of the rotor winding. The resulting magnetic field of the amortisseur (squirrel-cage) winding reacts with the stator field to create a torque, which causes the rotor to turn.

The rotation of the rotor will increase in speed to a point slightly below the synchronous speed of the stator—about 92 percent to 97 percent of the motor rated speed. There is a small slip in the speed of the rotor behind the speed of the magnetic field set up by the stator.

The field circuit is now excited from an outside source of direct current, and fixed magnetic poles are set up in the rotor field cores. The magnetic poles of the rotor are attracted to unlike magnetic poles of the magnetic field set up by the stator. Figures 8–1 and 8–2 show how the rotor field poles lock with unlike poles of the stator field. Once the field poles are locked, the rotor speed becomes the same as the speed of the magnetic field set up by the stator windings. In other words, the speed of the rotor is now equal to the synchronous speed.

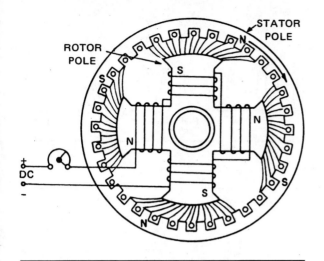

Figure 8–1   Diagram to show the principle of operation of a synchronous motor; Courtesy Alerich, *Electricity 4*, Delmar Publishers.

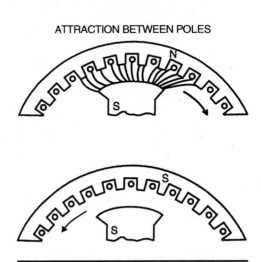

Figure 8–2   Starting of synchronous motor; Courtesy Alerich, *Electricity 4*, Delmar Publishers.

Remember that a synchronous motor must always be started as a three-phase squirrel-cage induction motor with the DC field excitation disconnected. The DC field circuit is added only after the rotor accelerates to a value near the synchronous speed. The motor will then operate as a synchronous motor, locked in step with the stator rotating field.

If an attempt is made to start a three-phase synchronous motor by first energizing the DC field circuit and then applying the three-phase voltage to the stator windings, the motor will not start, since the net torque is zero. At the instant the three-phase voltage is applied to the stator windings, the magnetic field set up by the stator current turns at the synchronous speed. The rotor, with its magnetic poles of fixed polarity, is attracted first by an unlike stator pole. It attempts to turn in that direction. However, before the rotor can turn, another stator pole of opposite polarity moves into position and the rotor then attempts to turn in the opposite direction. Because of this action of the alternating poles, the net torque is zero and the motor does not start.

Direct-Current Field Excitation

There are two methods to excite the synchronous motor. The first method is to excite the rotor with direct current. This is usually accomplished with a brushless exciter or DC generator attached to the shaft of the rotor. The stator windings are energized with three-phase power; the motor will run to just below synchronized when the DC excitation occurs. The rotor is energized with DC voltage and increases the speed of the rotor until the magnetic fields of the stator are in step with the rotor. Figure 8–3 shows the connections for a synchronous motor. A field rheostat in the separately excited field circuit varies the current in the field circuit. Changes in the field current affect the strength of the magnetic field developed by the revolving rotor. Variations in the rotor field strength do not affect the motor, which continues to operate at a constant speed. However, changes in the DC field excitation do change the power factor of a synchronous motor.

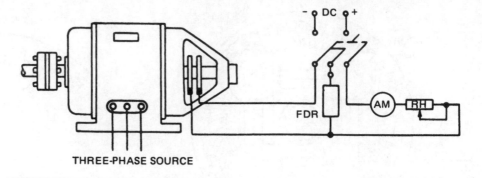

Figure 8–3 External connections for a synchronous motor; Courtesy Alerich, *Electricity 4,* Delmar Publishers.

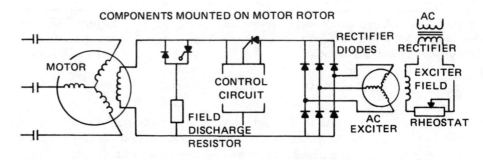

Figure 8–4 Simplified circuit for a brushless synchronous motor; Courtesy Alerich, *Electricity 4,* Delmar Publishers.

Brushless Solid-State Excitation

The second method of excitation is brushless solid state. This is a better method to provide excitation. The commutator of a conventional direct-connected exciter is replaced with a three-phase, bridge-type, solid-state rectifier. The DC output is then fed directly to the motor field winding. Simplified circuitry is shown in Figure 8–4. Color-shaded components are mounted on the rotor. A stationary field ring for the AC exciter receives DC from a small rectifier in the motor control cabinet. This rectifier is powered from the AC source. The exciter DC field is also adjustable. Rectifier solid-state diodes change the exciter AC output to DC. This DC is the source of excitation for the rotor field poles. Silicon-controlled rectifiers, activated by the solid-state field control circuit, replace electromechanical relays and the contactors of the conventional brush-type synchronous motor.

The field discharge resistor is inserted during motor starting. At motor synchronizing pull-in speed, the field discharge circuit is automatically opened and DC excitation is applied to the rotor field pole windings. Excitation is automatically removed if the motor pulls out of step (synchronization) due to an overload or a voltage failure. The stator of a brushless motor is similar to that of a brush-type motor. Mounted on the rotor shaft are the armature of the AC exciter, the AC output of which is rectified to DC by the silicon diodes. Brush and commutator problems are eliminated with this system.

Power Factor

A poor lagging power factor results when the field current is decreased below normal by inserting all of the resistance of the rheostat in the field circuit. The three-phase AC circuit to the stator supplies a magnetizing current that helps strengthen the weak DC field. This magnetizing component of current lags the voltage by 90 electrical degrees. Since the magnetizing component of current becomes a large part of the total current input, a low lagging power factor results.

If a weak DC field is strengthened, the power factor improves. As a result, the three-phase AC circuit to the stator supplies less magnetizing current. The magnetizing component of current becomes a smaller part of the total current input to the stator winding, and the power factor increases. If the field strength is increased sufficiently, the power factor increases to unity, or 100 percent. When a power factor value of unity is reached, the three-phase AC circuit does not supply any current and the DC field circuit supplies all of the current necessary to maintain a strong rotor field. The value of DC field excitation required to achieve unity power factor is called normal field excitation.

If the magnetic field of the rotor is strengthened further by increasing the DC field current above the normal excitation value, the power factor decreases. However, the power factor is leading when the DC field is overexcited. The three-phase AC circuit feeding the stator winding delivers a demagnetizing component of current that opposes the too-strong rotor field. This action results in a weakening of the rotor field to its normal magnetic strength.

The diagrams in Figure 8–5 show how the DC field is aided or opposed by the magnetic field set up by the AC windings. It is assumed in Figure 8–5 that the DC field is

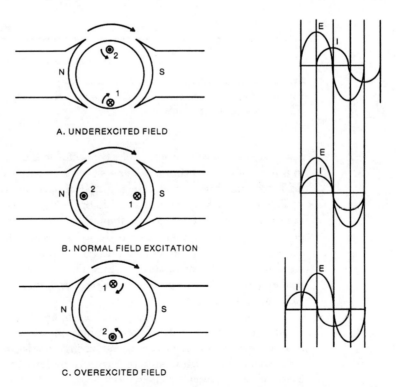

A. UNDEREXCITED FIELD

B. NORMAL FIELD EXCITATION

C. OVEREXCITED FIELD

Figure 8–5 Field excitation in a synchronous motor; Courtesy Alerich, *Electricity 4,* Delmar Publishers.

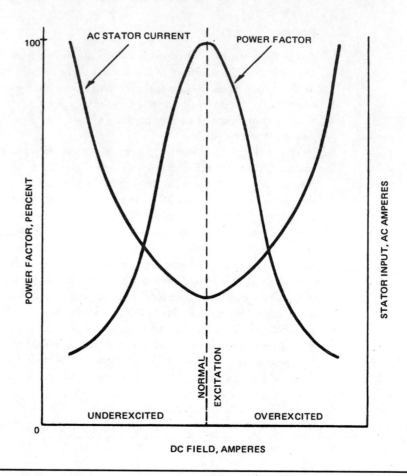

Figure 8–6 Characteristic operating curves for synchronous motors; Courtesy Alerich, *Electricity 4*, Delmar Publishers.

stationary and a revolving armature is connected to the AC source. The reader should keep in mind that most synchronous motors have stationary AC windings and a revolving DC field. For either case, however, the principle of operation is the same.

Figure 8–6 shows two characteristic operating curves for a three-phase synchronous motor. With normal full-field excitation, the power factor has a peak value of unity, or 100 percent, and the AC stator current is at its lowest value. As the DC field current is decreased in value, the power factor decreases in the lag quadrant and there is a resulting rapid rise in the AC stator current. If the DC field current is increased above the normal field excitation value, the power factor decreases in the lead quadrant and a rapid rise in the AC stator current results.

It has been shown that a synchronous motor operated with an overexcited DC field has a leading power factor. For this reason, a three-phase synchronous motor often is connected to a three-phase industrial feeder circuit having a low lagging power factor (Figure 8–7). In other words, the synchronous motor with an overexcited DC field will correct the power factor of the industrial feeder circuit.

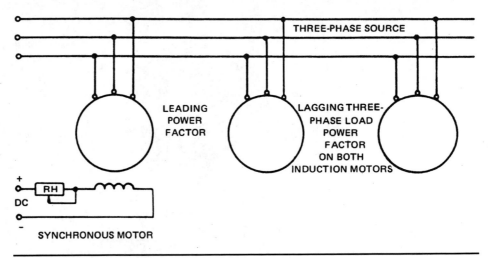

Figure 8–7 Synchronous motor used to correct power factor; Courtesy Alerich, *Electricity 4*, Delmar Publishers.

In Figure 8–7, two induction motors with lagging power factors are connected to an industrial feeder circuit. The synchronous motor connected to the same feeder is operated with an overexcited DC field. Since the synchronous motor can be adjusted so that the resulting power factor is leading, the power factor of the industrial feeder can be corrected until it reaches a value near unity, or 100 percent.

Reversing Rotation

The direction of rotation of a synchronous motor is reversed by interchanging any two of the three-line leads feeding the stator winding. The electrician must remember that the direction of rotation of the motor does not change if the two conductors of the DC source are interchanged.

Industrial Applications

The three-phase synchronous motor is used when a prime mover having a constant speed from a no-load condition to full load is required, such as fans, air compressors, and pumps. The synchronous motor is used in some industrial applications to drive a mechanical load and also to correct power factor. In some applications, this type of motor is used only to correct the power factor of an industrial power system. When the synchronous motor is used only to correct power factor and does not drive any mechanical load, it serves the same purpose as a bank of capacitors used for power factor correction. Therefore, in such an installation the motor is called a synchronous capacitor.

Three-phase synchronous motors, up to a rating of 10 horsepower, usually are started directly across the rated three-phase voltage. Synchronous motors of larger sizes are started through a starting compensator or an automatic starter. In this type of starting, the voltage applied to the motor terminals at the instant of start is about half the value of the rated line voltage, and the starting surge of current is limited.

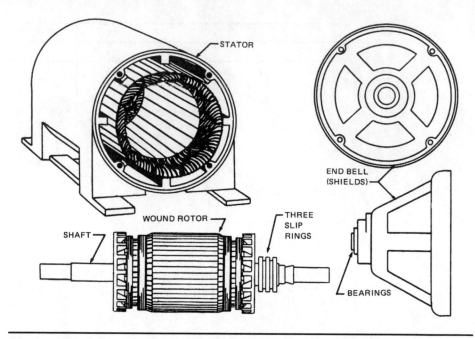

Figure 8–8 Parts of a wound-rotor motor; Courtesy Alerich, *Electricity 4*, Delmar Publishers.

Wound-Rotor Motor

Many industrial motor applications require three-phase motors with variable speed control. The squirrel-cage induction motor cannot be used for variable speed work, since its synchronous speed is essentially constant. Therefore, another type of induction motor was developed for variable speed applications. This motor is called the wound-rotor induction motor or slip-ring AC motor.

Construction Details

A three-phase, wound-rotor induction motor consists of a stator core with a three-phase winding, a wound rotor with slip rings, brushes, and brush holders, and two end shields to house the bearings that support the rotor shaft.

Figure 8–8 shows the basic parts of a three-phase, wound-rotor induction motor.

The Stator   A typical stator contains a three-phase winding held in place in the slots of a laminated steel core. The winding consists of formed coils arranged and connected so that there are three single-phase windings spaced 120 electrical degrees apart. The separate single-phase windings are connected either in wye or delta. Three line leads are brought out to a terminal box mounted on the frame of the motor. This is the same construction as the squirrel-cage motor stator.

The Rotor   The rotor consists of a cylindrical core composed of steel laminations. Slots cut into the cylindrical core hold the formed coils of wire for the rotor winding. The rotor winding consists of three single-phase windings spaced 120 electrical degrees apart. The single-phase windings are connected either in wye or delta. (The rotor winding must have the same number of poles as the stator winding.) The three leads from the three-phase rotor winding terminate at three slip rings mounted on the rotor shaft. Leads from carbon brushes that ride on these slip rings are connected to an external speed controller to vary the rotor resistance for speed control. More detail of this application is located in the "Motor Controls" volume of the *Electrician's Technical Reference*.

The brushes are held securely to the slip rings of the wound rotor by adjustable springs mounted in the brush holders. The brush holders are fixed in one position. For this type of motor, it is not necessary to shift the brush position as is sometimes required in direct-current generator and motor work.

The Motor Frame The motor frame is made of cast steel. The stator core is pressed directly into the frame. Two end shields are bolted to the cast steel frame. One of the end shields is larger than the other because it must house the brush holders and brushes that ride on the slip rings of the wound rotor. In addition, it often contains removable inspection covers.

The bearing arrangement is the same as that used in squirrel-cage induction motors. Either sleeve bearings or ball-bearing units are used in the end shields.

Principles of Operation

When three currents, 120 electrical degrees apart, pass through the three single-phase windings in the slots of the stator core, a rotating magnetic field is developed. This field travels around the stator. The speed of the rotating field depends on the number of stator poles and the frequency of the power source. This speed is called the synchronous speed. It is determined by applying the formula that was used to find the synchronous speed of the rotating field of squirrel-cage induction motors.

$$\text{Synchronous speed in r / min} = \frac{120 \times \text{frequency in hertz}}{\text{number of poles}}$$

$$S = \frac{120 \times f}{P}$$

As the rotating field travels at synchronous speed, it cuts the three-phase winding of the rotor and induces voltages in this winding. The rotor winding is connected to the three slip rings mounted on the rotor shaft. The brushes riding on the slip rings connect to an external wye-connected group of resistors (speed controller), as shown in Figure 8–9. The induced voltages in the rotor windings set up currents that follow a closed path from the rotor winding to the wye-connected speed controller. The rotor currents create a magnetic field in the rotor core based on transformer action. This rotor field reacts with the stator field to develop the torque, which causes the rotor to turn. The speed controller is sometimes called the secondary resistance control.

Starting Theory of Wound-Rotor Induction Motors

To start the motor, all of the resistance of the wye-connected speed controller is inserted in the rotor circuit. The stator circuit is energized from the three-phase line. The voltage induced in the rotor develops currents in the rotor circuit. The rotor currents, however, are limited in value by the resistance of the speed controller. As a result, the stator current also is limited in value. In other words, to minimize the starting surge of current to a wound-rotor induction motor, insert the full resistance of the speed controller in the rotor circuit. As the motor accelerates, steps of resistance in the wye-connected speed controller can be cut out of the rotor circuit until the motor accelerates to its rated speed.

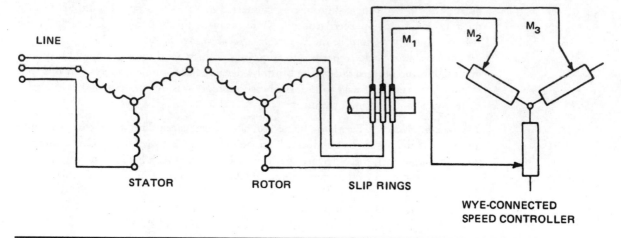

LINE

STATOR   ROTOR   SLIP RINGS

M_1 M_2 M_3

WYE-CONNECTED
SPEED CONTROLLER

Figure 8–9 Connections for a wound-rotor induction motor and a speed controller; Courtesy Alerich, *Electricity 4,* Delmar Publishers.

Speed Control

The insertion of resistance in the rotor circuit not only limits the starting surge of current, but also produces a high starting torque and provides a means of adjusting the speed. If the full resistance of the speed controller is cut into the rotor circuit when the motor is running, the rotor current decreases and the motor slows down. As the rotor speed decreases, more voltage is induced in the rotor windings and more rotor current is developed to create the necessary torque at the reduced speed.

If all of the resistance is removed from the rotor circuit, the current and the motor speed will increase. However, the rotor speed will always be less than the synchronous speed of the field developed by the stator windings. Recall that this fact also is true of the squirrel-cage induction motor. The speed of a wound-rotor motor can be controlled manually or automatically with timing relays, contactors, and pushbutton speed selection. Control techniques are given in detail in the "Motor Controls" volume of the *Electrician's Technical Reference.*

Torque Performance

As a load is applied to the motor, both the percent slip of the rotor and the torque developed in the rotor increase. As shown in the graph in Figure 8–10, the relationship between the torque and percent slip is practically a straight line.

Figure 8–10 illustrates that the torque performance of a wound-rotor induction motor is good whenever the full resistance of the speed controller is inserted in the rotor circuit. The large amount of resistance in the rotor circuit causes the rotor current to be almost in phase with the induced voltage of the rotor. As a result, the field set up by the rotor current is almost in phase with the stator field. If the two fields reach a maximum value at the same instant, there will be a strong magnetic reaction resulting in a high torque output. However, if all of the speed controller resistance is removed from the rotor circuit and the motor is started, the torque performance is poor. The rotor circuit minus the speed controller resistance consists largely of inductive reactance. This means that the rotor current

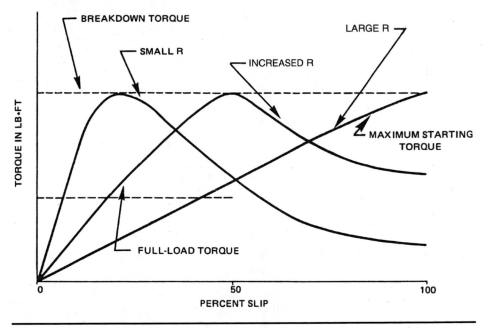

Figure 8–10 Performance curves of a wound-rotor motor; Courtesy Alerich, *Electricity 4*, Delmar Publishers.

lags behind the induced voltage of the rotor, and thus, the rotor current lags behind the stator current. As a result, the rotor field set up by the rotor current lags behind the stator field, which is set up by the stator current. The resulting magnetic reaction of the two fields is relatively small, since they reach their maximum values at different points. In summary, then, the starting torque output of a wound-rotor induction motor is poor when all resistance is removed from the rotor circuit.

Speed Regulation

It was shown in the previous paragraphs that the insertion of resistance at the speed controller improves the starting torque of a wound-rotor motor at low speeds. However, there is an opposite effect at normal speeds. In other words, the speed regulation of the motor is poorer when resistance is added in the rotor circuit at a higher speed. For this reason, the resistance of the speed controller is removed as the motor comes up to its rated speed.

Figure 8–11 shows the speed performance of a wound-rotor induction motor. Note that the speed characteristic curve resulting when all of the resistance is cut out of the speed controller indicates relatively good speed regulation. The second speed characteristic curve, resulting when all of the resistance is inserted in the speed controller, has a marked drop in speed as the load increases. This indicates poor speed regulation.

Power Factor

The power factor of a wound-rotor induction motor at no load is as low as 15 percent to 20 percent lag. However, as the load is applied to the motor, the power factor improves and increases to 85 percent to 90 percent lag at rated load.

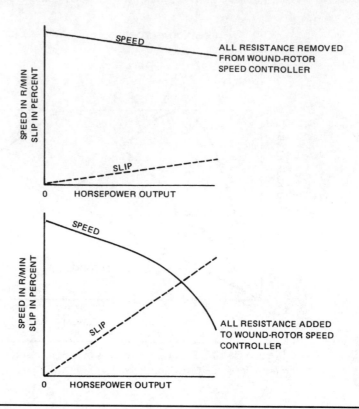

Figure 8–11   Speed performance curves of a wound-rotor motor; Courtesy Alerich, *Electricity 4,* Delmar Publishers.

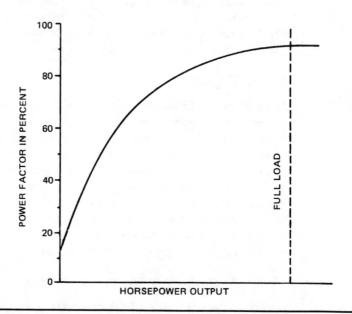

Figure 8–12   Power factor curve of a wound-rotor induction motor; Courtesy Alerich, *Electricity 4,* Delmar Publishers.

Figure 8–12 is a graph of the power factor performance of a wound-rotor induction motor from a no-load condition to full load. The low lagging power factor at no load is due to the fact that the magnetizing component of load current is such a large part of the total motor current. The magnetizing component of load current magnetizes the iron causing interaction between the rotor and the stator, by mutual inductance.

As the mechanical load on the motor increases, the in-phase component of current increases to supply the increased power demands. The magnetizing component of the current remains the same, however. Since the total motor current is now more nearly in phase with the line voltage, there is an improvement in the power factor.

Operating Efficiency

Both a wound-rotor induction motor with all of the resistance cut out of the speed controller and a squirrel-cage induction motor show nearly the same efficiency performance. However, when a motor must operate at slow speeds with all of the resistance cut in the rotor circuit, the efficiency of the motor is poor because of the power loss in watts in the resistors of the speed controller.

Figure 8–13 illustrates the efficiency performance of a wound-rotor induction motor. The upper curve showing the highest operating efficiency results when the speed controller is in the fast position and there is no resistance inserted in the rotor circuit. The lower curve shows a lower operating efficiency. This occurs when the speed controller is in the slow position and all of the controller resistance is inserted in the rotor circuit.

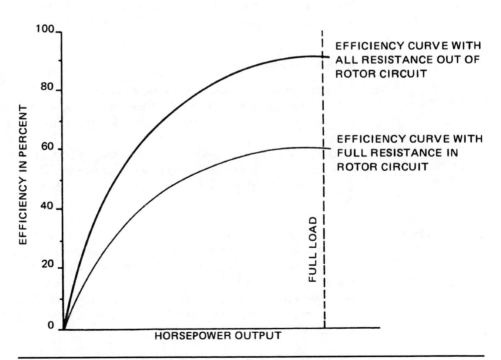

Figure 8–13   Efficiency curves for a wound-rotor induction motor; Courtesy Alerich, *Electricity 4,* Delmar Publishers.

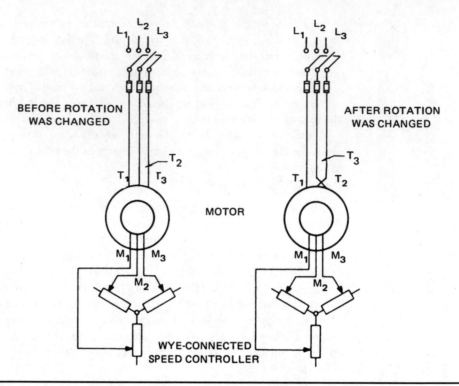

BEFORE ROTATION
WAS CHANGED

AFTER ROTATION
WAS CHANGED

MOTOR

WYE-CONNECTED
SPEED CONTROLLER

Figure 8–14 Changes necessary to reverse direction of rotation of a wound-rotor motor; Courtesy Alerich, *Electricity 4,* Delmar Publishers.

Reversing Rotation

The direction of rotation of a wound-rotor induction motor is reversed by interchanging the connections of any two of the three line leads (Figure 8–14). This procedure is identical to the procedure used to reverse the direction of rotation of a squirrel-cage induction motor.

The electrician should never attempt to reverse the direction of rotation of a wound-rotor induction motor by interchanging any of the leads feeding from the slip rings to the speed controller. Changes in these connections will not reverse the direction of rotation of the motor.

Selsyn Units

The word *selsyn* is an abbreviation of the words *self-synchronous.* These units are also referred to as *synchros,* and are known by various trade names. Synchros are primarily used for their ability to handle special torque applications such as control and remote signaling, and as indicating devices. These motors are used primarily in conjunction with other motors for control purposes. They are available for three-phase applications, which makes them popular. Synchros are detailed in the "Motor Controls" volume of the *Electrician's Technical Reference.*

Operation Standard Selsyn System

A selsyn system consists of two three-phase induction motors. The normally stationary rotors of these induction motors are interconnected so that a manual shift in the rotor posi-

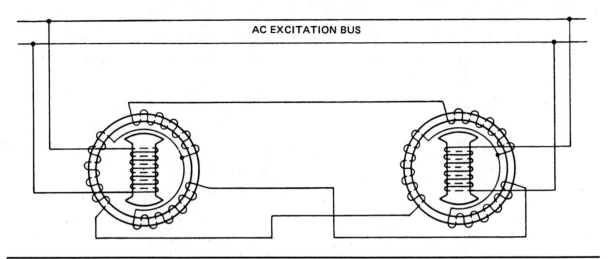

Figure 8–15   Diagram of selsyn motors showing interconnected stator and rotor windings connected to excitation source; Courtesy Alerich, *Electricity 4*, Delmar Publishers.

tion of one machine is accompanied by an electrical rotor shift in the other machine in the same direction and of the same angular displacement as the first unit.

Figure 8–15 shows a simple selsyn system for which the units at the transmitter and receiver are identical. The rotors of these units are two pole and must be excited from the same AC source. The three-phase stator windings are connected to each other by three leads between the transmitter and the receiver units. The rotor of each machine is called the primary, and the three-phase stator winding of each machine is called the secondary. A rotor for a typical selsyn unit is shown in Figure 8–15.

When the primary excitation circuit is closed, an AC voltage is impressed on the transmitter and receiver primaries. If both rotors are in the same position with respect to their stators, no movement occurs. If the rotors are not in the same relative position, the freely movable receiver rotor will turn to assume the same position as the transmitter rotor.

If the transmitter rotor is turned, either manually or mechanically, the receiver rotor will follow at the same speed and in the same direction.

The self-synchronous alignment of the rotors is the result of voltages induced in the secondary windings. Both rotors induce voltages into the three windings of their stators. These voltages vary with the position of the rotors. If the two rotors are in the same relative position, the voltages induced in the transmitter and receiver secondaries will be equal and opposite. In this condition, current will not exist in any part of the secondary circuit.

If the transmitter rotor is moved to another position, the induced voltages of the secondaries are no longer equal and opposite and currents are present in the windings. These currents establish a torque that tends to return the rotors to a synchronous position. Since the receiver rotor is free to move, it makes the adjustment. Any movement of the transmitter rotor is accompanied immediately by an identical movement of the receiver rotor.

Differential Selsyn System

Figure 8–16 is a diagram of the connections of a differential selsyn system consisting of a transmitter, a receiver, and a differential unit. This system produces an angular indication of the receiver. The indication is either the sum or difference of the angles existing at the

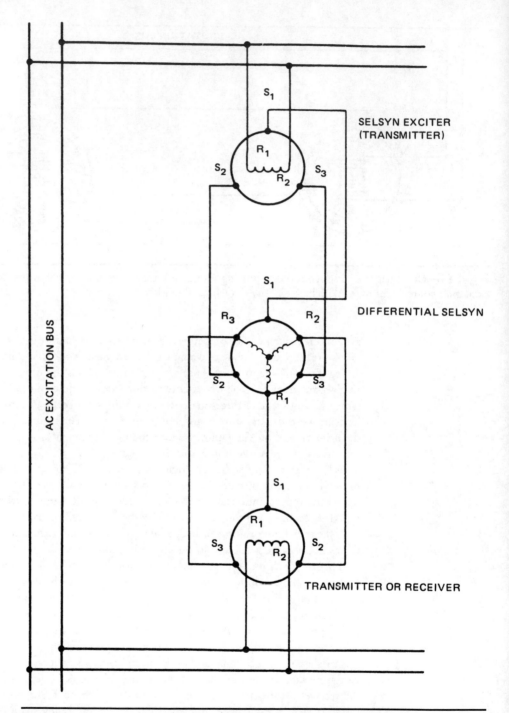

Figure 8–16   Schematic diagram of differential selsyn connections; Courtesy Alerich, *Electricity 4*, Delmar Publishers.

transmitter and differential selsyns. If two selsyn generators, connected through a differential selsyn, are moved manually to different angles, the differential selsyn will indicate the sum or difference of their angles.

A differential selsyn has a primary winding with three terminals. Otherwise, it closely resembles a standard selsyn unit. The three primary leads of the differential selsyn are brought out to collector rings. The unit has the appearance of a miniature wound-rotor, three-phase induction motor. The unit, however, normally operates as a single-phase transformer.

The voltage distribution in the primary winding of the differential selsyn is the same as that in the secondary winding of the selsyn exciter. If any one of the units is fixed in position and a second unit is displaced by a given angle, then the third unit that is free to rotate will turn through the same angle. The direction of rotation can be reversed by interchanging any pair of leads on either the rotor or stator winding of the differential selsyn.

If any two of the selsyns are rotated simultaneously, the third selsyn will turn through an angle equal to the algebraic sum of the movements of the two selsyns. The algebraic sign of this value depends on the direction of rotation of the rotors of the two selsyns, as well as the phase rotation of their windings.

The excitation current of the differential selsyn is supplied through connections to one or both of the standard selsyns to which the differential selsyn is connected. In general, the excitation current is supplied to the primary winding only. In this case, the selsyn connected to the differential stator supplies this current and must be able to carry the extra load without overheating. A particular type of selsyn, known as an exciter selsyn, is used to supply the current. The exciter selsyn can function in the system either as a transmitter or a receiver.

Advantages of Selsyn Units

Selsyn units are compact and rugged and provide accurate and very reliable readings. Because of the comparatively high torque of the selsyn unit, the indicating pointer does not oscillate as it swings into position. Internal mechanical dampers are used in selsyn receivers to prevent oscillation during the synchronizing procedure and to reduce any tendency of the receiver to operate as a rotor. The operation of the receiver is smooth and continuous and is in agreement with the transmitter. In addition, the receiver rapidly responds to changes in position at the transmitter.

In the event of a power failure, the indicator of the receiver resets automatically with the transmitter when power is received. Calibration and time-consuming checks are unnecessary. Selsyn systems offer several advantages:

- The indicators are small and compact and can be located where needed.
- The simple installation requires running a few wires and bolting the selsyn units in place.
- Selsyn units can be used to indicate either angular or linear movement.
- Selsyn units control the motion of a device at a distant point by controlling its actuating mechanism.
- One transmitter may be used to operate several receivers simultaneously at several distant points.

CHAPTER 9

Motor Testing Procedures and Techniques*

AC Motors

Visual and Mechanical Inspection

- Inspect for physical damage.
- Inspect for proper anchorage, mounting, grounding, connection, and lubrication.
- When applicable, perform special tests such as air gap spacing and pedestal alignment.

Electrical Tests—Induction Motors

Perform insulation-resistance tests in accordance with ANSI/IEEE Standard 43.

Motors Larger than 200 Horsepower Test duration shall be for 10 minutes with resistances tabulated at 30 seconds, one 1 minute, and 10 minutes. Calculate polarization index using the following procedure.

1. Apply either 500 or 1000 volts DC between the winding and ground using a direct-indicating, power-driven megohm-meter. For machines rated 500 volts and over, the higher value is used.

2. Apply voltage for 10 minutes and keep constant for the duration of the test. The polarization index is calculated as the ratio of the 10-minute to the 1-minute value of the insulation resistance, measured consecutively:

$$\text{Polarization index} = \frac{\text{Resistance after 10 minutes}}{\text{Resistance after 1 minute}}$$

3. The recommended minimum value of polarization index for AC and DC motors is 2.0. Machines having windings with a lower index are less likely to be suited for operation.

*Adapted from (NETA) National Electrical Testing Association, testing specifications standard.

The **polarization index** is useful in evaluating windings for the following:

- Buildup of dirt or moisture
- Gradual deterioration of the insulation (by comparing results of tests made earlier on the same machine)
- Fitness for overpotential tests
- Suitability for operation

Motors 200 Horsepower and Less   Test duration shall be for 1 minute with resistances tabulated at 30 and 60 seconds. Calculate the dielectric absorption ratio. Perform DC overpotential tests on motors rated at 1000 horsepower and greater and at 4000 volts and greater in accordance with ANSI/IEEE Standard 95.

- Perform insulation power-factor or dissipation-factor tests.
- Perform surge comparison tests.
- Perform insulation–resistance test on pedestal per manufacturer's instructions.
- Perform insulation–resistance test on surge protection device.
- Test motor starter in accordance with manufacturer's specifications prior to reenergizing the motor.
- Check resistance temperature detector (RTD) circuits for conformance with drawings.
- Check that metering or relaying devices using the RTDs are of the proper rating.
- Check that the motor space heater is operating.
- Perform a rotation test to ensure proper shaft direction if the motor has been electrically disconnected.
- Measure running current and evaluate relative to load conditions and nameplate full-load amperes.

Perform Vibration Tests
1. Motors larger than 200 horsepower: Perform vibration baseline test. Amplitude shall be plotted versus frequency.
2. Motors 200 horsepower and less: Perform vibration and amplitude test.

Electrical Tests—Synchronous Motors

- Perform all tests as indicated in "Induction Motor Test" section of this chapter.
- Perform a voltage-drop test on all salient poles.
- Perform insulation-resistance tests on the main rotating field winding, the exciter field winding, and the exciter armature winding in accordance with ANSI/IEEE Standard 43.
- Perform a high-potential test on the excitation system in accordance with ANSI/IEEE Standard 421B.
- Measure and record resistance of motor-field winding, exciter-stator winding, exciter rotor windings, and field discharge resistors.

- Perform front-to-back resistance tests on diodes and gating tests of SCR's for field application semiconductors.
- Check that the field application and power-factor relay (enable time-delay relays) have been tested and set to the motor drive manufacturer's recommended values.
- Record stator current, stator voltage, and field current by strip chart recorder for the complete acceleration period including stabilization time for a normally loaded starting condition. From the recording, determine the following information:
  1. Time to reach stable synchronous operation.
  2. Bus voltage prior to start.
  3. Voltage drop at start.
  4. Bus voltage at motor full-load.
  5. Locked-rotor current.
  6. Current after synchronization but before loading.
  7. Full-load current.
  8. Acceleration time to near synchronous speed.
  9. RPM just prior to synchronization.
  10. Field application time.
- Plot a V-curve of stator current versus excitation current at approximately 50 percent (50%) load to check proper exciter operation.
- If range of exciter adjustment and motor loading permit, reduce excitation to cause power factor to fall below the trip value of the power-factor relay. Check relay operation.

Test Values
- Insulation–resistance test results should comply with values listed in *Table A* in the appendix.
- Investigate dielectric absorption ratios less than 1.4 and polarization index ratios less than 1.5 for Class A insulation and 2.0 for Class B insulation.
    Note: Overpotential, high-potential, and surge comparison tests shall not be made on motors having values lower than those just indicated.
- Stator winding DC overpotential test voltage shall be in accordance with NEMA publication MG-1, paragraph 3.01.L. Test results are dependent on ambient conditions, and evaluation is on a withstand basis. If phase windings can be separately tested, values of leakage current may be compared for similar windings.
- Vibration amplitudes shall not exceed values shown in Chapter 9.
- Salient pole voltage drop should be equal for each pole. Investigate values that differ by more than 10 percent (10%).

The measured resistance values of motor-field winding, exciter-stator winding, exciter rotor windings, and field discharge resistors shall be compared to manufacturer's recommended values.

DC Motors
Visual and Mechanical Inspection

- Perform inspection and tests as outlined in section for AC motors.
- Inspect brushes and brush rigging.

Electrical Tests

Perform insulation-resistance tests on all windings in accordance with ANSI/IEEE Standard 43.

Motors Larger than 200 Horsepower   Test duration shall be for 10 minutes with resistances tabulated at 30 seconds, 1 minute, and 10 minutes. Calculate polarization index using the following procedure.

1. Apply either 500 or 1000 volts DC between the winding and ground.
2. Apply voltage for 10 minutes and keep constant for the duration of the test. The polarization index is calculated as the ratio of the 10-minute to the 1-minute value of the insulation resistance, measured consecutively:

$$\text{Polarization index} = \frac{\text{Resistance after 10 minutes}}{\text{Resistance after 1 minute}}$$

3. The recommended minimum value of polarization index for AC and DC motors is 2.0. Machines having windings with a lower index are less likely to be suited for operation.

The polarization index is useful in evaluating windings for:

1. Buildup of dirt or moisture.
2. Gradual deterioration of the insulation (by comparing results of tests made earlier on the same machine).
3. Fitness for overpotential tests.
4. Suitability for operation.

Motors 200 Horsepower and Less   Test duration shall be for 1 minute with resistances tabulated at 30 and 60 seconds. Calculate the dielectric absorption ratio. Perform high-potential test in accordance with NEMA publication MG-1, paragraph 3.01.

- Perform a voltage-drop test on all field poles.
- Perform insulation power-factor or dissipation-factor tests.
- Measure armature running current and field current or voltage. Compare to nameplate.

Perform Vibration Tests
1. Motors larger than 200 horsepower: Perform vibration baseline test. Amplitude to be plotted versus frequency.
2. Motors 200 horsepower and less: Perform vibration amplitude test.

Check all protective devices in accordance with other sections of these specifications.

Test Values
1. Insulation resistance test values shall comply with those listed in *Table A* in the appendix. Investigate dielectric absorption ratios less than 1.4 and polarization index ratios less than 1.5 for Class A insulation and 2.0 for Class B insulation.
    Note: Overpotential, high-potential, and surge comparison tests shall not be made on motors having values lower than those already indicated.

2. Overpotential test evaluation shall be on a withstand basis.

3. Vibration amplitudes shall not exceed the following.

Maximum Allowable Vibration Amplitude	
Speed–RPM	*Amplitude–Inches Peak to Peak*
• 3000 and above	.001
• 1500–2999	.002
• 1000–1499	.0025
• 999 and below	.003

CHAPTER

10

Tuning Your Motor System For Industrial And Commercial Application*

Just as an eight-cylinder automobile must have all eight cylinders operating in sync, so must the motor(s) of a plant or machine to operate properly. Without proper operation, the machine or plant will not operate efficiently. Proper operation of the motor system is essential if we are to obtain the best production from our machine or plant. Simply put, if our machine or plant is operating at 80 percent of its capacity then we are suffering a 20 percent lost. This may not be important during a time of low production, but it is costly when the market calls for full production. If your plant is to operate at maximum potential, tuning your motors to greatest efficiency is a must. In this chapter you will be given tips, techniques, tables, charts, graphs, and calculations to make your motor system operate at maximum potential and reduce operational cost. Proper application of motors to its load is important to obtain maximum potential.

The industrial plant electrician is identified as the person who can be most effective in increasing electrical energy efficiency in the plant. However, in order to do so, the electrician needs some basic measurement tools and an understanding of how induction motors relate to the in-plant electrical distribution system.

The Electrician's Role

Historically, the plant electrician has the responsibility for keeping motors in a plant in good operation. It was, and still is, common for a part of a manufacturing plant to be down because of a single electric motor failure. Under those circumstances, the plant electrician works quickly to replace the failed unit so there is minimal disruption to the plant operating schedule. Often, the best way to minimize the downtime is to have a replacement motor on hand, ready to install. However, it is too costly to have a spare in inventory for every motor in the plant. It is not particularly surprising, then, that cases arise where the proper size replacement motor is not in inventory. It is not unusual for the electrician to put the next larger size available into service.

Typically, the intention is to have the failed motor rebuilt and then put it back in service when time permits. However, there is rarely enough time to replace a temporary working

*Adapted from DOE Departmant of Energy, Motor Challenge Division.

motor, so the motor that is returned from the repair shop goes into inventory, waiting for another motor to fail. In this way, the size of motors at many industrial facilities grows over time. In older plants this increase can be considerable.

When properly instructed and equipped with the right tools, the industrial electrician can become the pivotal individual in the task of saving electrical energy through proper motor analysis and selection. Ideally, a plant electrician will develop a plan for upgrading motors. A well-conceived plan should be in place before the next motor failure occurs.

Rudiments of the plan call for an inventory of plant motors. Included in the inventory data base will be:

- Precise individual motor identification
- Accurate motor load (in kW) determination
- Current power factor determination
- Current motor efficiency determination
- Action to be taken at failure (i.e., repair or replacement specifications)

The electrician should make a list of motors to evaluate. Loads that are significant energy users should head the list. This means that large motors that operate for extended periods are most important. Conversely, small motors that run for short periods are least important. The objective is to sort through all the motors and prioritize the list using common sense.

Each motor on the list is to be analyzed for replacement on a size and efficiency basis. In order to carry out this task, the electrical staff will need to operate in an organized fashion. This *Electrician's Technical Reference* provides detailed instruction to assist in that effort.

Power Quality Considerations

If there are variable speed drives, induction heaters, or other electronic loads on the system, voltage harmonics can be expected. Extreme current harmonics are present in circuits feeding these loads. The purchaser should describe this electrical environment to the equipment supplier and ascertain the capability of alternative devices to accurately measure such "dirty" power.

At the minimum, devices that sense voltage or current must operate on a true RMS principle. Those that do not sense only the peak of the wave, and assume a sinusoidal shape, thus inferring RMS as 0.7071 times the peak. For power factor and power, instruments that are not true RMS determine phase lag by the time lapse between when the voltage and current waves cross zero.

Multi-Channel Power Loggers

Before undertaking the field measurements necessary to support a motor improvement program, the energy coordinator should identify and prioritize motors that are candidates for replacement with energy-efficient units. The in-plant electrical distribution system should be examined and "tuned" to eliminate over- and under-voltage and voltage unbalance prior to conducting motor field measurements.

Over- and Undervoltage

Usual service conditions, defined in the National Electrical Manufacturer's Association (NEMA) Standards Publication MG-1- 1993, Rev. 1, *Motors and Generators,* include operation within a tolerance of + 10 percent of a motors' rated voltage.

Overvoltage

If the voltage at the motor feeder is increased, the magnetizing current increases. At some point, depending on the design of the motor, saturation of the core iron will increase and overheating will occur. At about 10 to 15 percent overvoltage, both efficiency and power factor significantly decrease for standard efficiency motors while the full-load slip decreases. The starting current, locked rotor torque, and breakdown torque all significantly increase with overvoltage conditions.

A voltage that is at the high end of tolerance limits frequently indicates that a transformer tap has been moved in the wrong direction. An overload relay will not recognize this overvoltage situation, and, if the voltage is more than 10 percent high, the motor can overheat. Overvoltage operation with VAR currents above acceptable limits for extended periods of time can accelerate deterioration of a motor's insulation.

Reduced operating efficiency because of low voltages at the motor terminals is generally due to excessive voltage drops in the supply system. If the motor is at the end of a long feeder, reconfiguration may be necessary.

The system voltage can also be modified by:

- Adjusting the transformer tap settings
- Installing automatic tap-changing equipment if system loads vary considerably over the course of a day
- Installing power factor correction capacitors that raise the system voltage while correcting for power factor

Since motor efficiency and operating life are degraded by voltage variations, only motors with compatible voltage nameplate ratings should be specified for a system.

For example, three-phase motors are available with voltage ratings of 440, 460, 480, and 575 volts. The use of a motor designed for 460-volt service in a 480-volt system results in reduced efficiency, increased heating, and reduced motor life. A 440-volt motor would be even more seriously affected.

Undervoltage

If a motor is operated at reduced voltage, even within the allowable 10 percent limit, the motor will draw increased current to produce the torque requirements imposed by the load. This causes an increase in both stator and rotor I^2R losses. Low voltages can also prevent the motor from developing an adequate starting torque.

The custom by NEMA members is to rate motors at 95.8 percent of nominal system voltage. For example, motors intended for use on 480-volt systems are rated at 460 volts (95.8% × 480V) and motors intended or use on 240-volt systems are rated at 230 volts (95.8% × 240V). Motors can be allowed to operate on voltages as low as 95.6 percent of their specified voltage rating. Thus, a motor rated at 460 volts can operate at 440 volts (460V × 0.956). As long as the phase-to-phase voltages are balanced, the motor need not be derated.

In most industrial systems the nominal voltage value is 480 volts. Delta-type electrical systems always refer to "line to line" voltage values. Wye-type electrical systems refer to "line-to-line" and "line-to-neutral" voltage values. One will frequently see a voltages described as 277/480. This means that the voltage from the line-to-neutral is 277 volts and the voltage from line-to-line is 480 volts. More information on this subject is located in the "Power and Distribution" volume of the *Electricians' Technical Reference*.

The serving utility is obligated to deliver power to the 480 volt industrial user in the range from a low of 456 volts to a high of 504 volts (480 + 5%). In practice, the service voltage is usually maintained within a tight range. It is common to experience the service voltage remaining in a range of 475 volts to 485 volts. Acceptable system delivery voltage values are defined by state electrical codes.

Voltage Unbalance

A voltage unbalance occurs when there are unequal voltages on the lines to a polyphase induction motor. This unbalance in phase voltages causes the line currents to be out of balance. The unbalanced currents cause torque pulsations, vibrations, increased mechanical stress on the motor, and overheating of one and possibly two of the phase windings. Voltage unbalance has a detrimental effect on motors. Motor efficiency suffers when motors are subjected to significant voltage unbalances. When efficiency falls, energy drawn by the motor dissipates as heat in the core and in the windings. Useful torque at the shaft is reduced. Ultimately, the motor can fail from insulation breakdown. Figure 10–1 is a derating curve published by NEMA. The amount of derating for a motor is described by the curve as a function of voltage unbalance. Unbalance of more than 1 percent requires that a motor be derated and may void a manufacturer's warranty. NEMA recommends against using motors where the unbalance is greater than 5 percent. Voltage unbalance is defined by NEMA as 100 times the maximum deviation of the line voltage from the average voltage on a three-phase system, divided by the average voltage. Most utilities attempt to control the service voltage unbalance to less than 2 percent.

In a system where the three line-to-line or (phase to phase) service voltages are:

484.5 volts, 482.0 volts, and 469.7 volts, the average of the three is 478.7 volts.

Maximum deviation from average is 9.0 V (478.7 V – 469.7 V)

Percent voltage unbalance = (9.0 V) / (478.7 V) × 100 = 1.9%

With a well-designed electrical distribution system in the plant, the amount of unbalance at the load and motor control centers should be about the same as at the degree of unbalance at the service entrance. When the unbalance is significantly different at the load centers, there is a phase voltage drop problem between the power entrance panel and the load centers. The unbalancing problems must be found within the plant and must be corrected prior to recording motor data.

The effect of unbalance not only distorts potential energy savings, but it may cause irreparable damage to equipment.

Trouble-shooting and Tuning Your In-Plant Distribution System

Maintenance of in-plant electrical distribution systems is often neglected, with increased costs ultimately being paid in the forms of decreased safety (due to increased fire hazard), decreased motor life, increased downtime, and lost productivity.

Efficiency improvements can be achieved through eliminating such common problems as poor contacts, undersized conductors, voltage unbalance, over- and undervoltage, low power factor, and insulation leakage. These will be discussed later in this chapter. In-plant distribution system troubleshooting and correction should occur before field data measurements are taken.

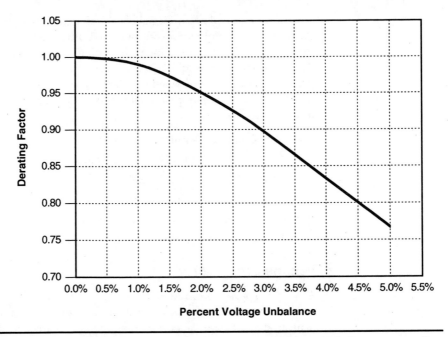

Figure 10–1   Motor voltage unbalance derating curve; Source: U.S. Department of Energy.

The electrical distribution system tune-up should begin with a search for and correction of poor contacts, since these are most likely to result in a catastrophic system failure and possibly a fire. Correction of poor power factor is the next step, since this is generally the source of the greatest utility cost savings. The system should then be examined for voltage unbalance and over/under voltage conditions due to their detrimental effect on motor performance and motor life. A survey for insulation leakage and undersized conductors should also be made.

Poor Connections

The first step in optimizing your industrial electrical distribution system is to detect and correct any problems due to poor connections. High temperatures are commonly caused by loose and dirty contacts. Such contacts are found in switches, circuit breakers, fuse clips, and terminations. These problems are the most cost-effective to correct.

Poor contacts can be caused by:

- Loose cable terminals and bus bar connections.
- Corroded terminals and connections
- Poor crimps or bad solder joints
- Loose, pitted, worn, or poorly adjusted contacts in motor controllers or circuit
- breakers

Fuse clips and breaker fingers are the major sources of poor connections.

Detection should begin with either an infrared thermography or voltage drop survey of the power panels and motor control centers. Advantages of a voltage drop survey are that the survey can be done in-house with existing equipment and that problems can often be detected before they would be with infrared thermography. Voltage drop measurements should be taken at a time when the plant is heavily loaded.

During the voltage drop or infrared survey, the electrician can visually inspect suspected trouble areas for:

- Discoloration of insulation or contacts
- Compromised insulation ranging from small cracks to bare conductors
- Oxidation of conductor metals
- Presence of contaminants such as dirt
- Mismatched cables in common circuits
- Aluminum cables connected to lugs marked for copper wire

Voltage Drop Survey

A *voltage drop survey* can be done with a simple hand-held milli-voltmeter. A two-stage process can be used to quickly identify problems without individually testing each component. In the first stage, voltage drop measurements are taken from the input of each panel to the panel output to each load. An example for a typical motor circuit is to measure the voltage drop from the bus bar to the load side of the motor starter. Comparing the magnitude of voltage drop with other phases supplying the load can alert the electrician to poor connections. For the second stage of the survey, the electrician can make component-by-component voltage drop measurements on suspect circuits to isolate and eliminate poor connections.

Infrared thermography is a quick and reliable method for identifying and measuring temperatures of components operating at unreasonably elevated temperatures. High temperatures are a strong indication of both energy wastage and pending failure. High-resistance connections are self-aggravating since they generate high temperatures, which further reduce component conductivity and increase the operating temperature.

Once the infrared survey is complete, the plant electrician can focus on the located hot spots. A milli-voltmeter or milli-ohmmeter is recommended for measuring the voltage drop and resistance, respectively, across high-temperature connections and connections not shown on the thermographs. Use these measurements to determine if the hot spot indicated in the infrared image is due to a problem with the component itself or if heat is being radiated from an adjacent source.

Voltage Unbalance

Further efforts to optimize your electrical distribution system should include a survey of loads to detect voltage unbalances. Unbalances in excess of 1 percent should be corrected as soon as possible. A voltage unbalance of less than 1 percent is satisfactory. Figure 10–2 displays motor loss increases caused by voltage unbalance.

The following causes of voltage unbalance can be detected by sampling the voltage balance at a few locations:

- Selection of wrong taps on the distribution transformer or unequal tap settings, usually 1.25 or 2.5 percent per tap

- Presence of a large, single-phase distribution transformer on a polyphase system, whether it is under load or not
- Asymmetrical (unbalanced) transformer windings delivering different voltages
- Faulty operation of automatic equipment for power factor correction
- Unbalanced three-phase loads (such as lighting or welding)
- Single-phase loads unevenly distributed on a polyphase system, or large single-phase load connected to two conductors on a three-phase system
- Well-intentioned changes, such as improvements in the efficiency of single-phase lighting loads, which inadvertently bring a previously balanced polyphase supply into unbalance (possibly wasting more energy than was saved)
- Highly reactive single-phase loads, such as welders.
- Irregular on/off cycles of large loads such as arc furnaces or major banks of lights
- Unbalanced or unstable polyphase supply from the grid.

The following problems may be more critical in nature, resulting in two-phase or single-phase operation:

- An open phase on the primary side of a three-phase transformer in the distribution system
- Single phase-to-ground faults
- Failure or disconnection of one transformer in a three-phase delta-connected bank
- Faults, usually to ground in the power transformer
- A blown fuse or other open circuit on one or two-phases of a three-phase bank of power-factor correction capacitors
- Certain kinds of a single-phase failures in adjustable frequency drives and other motor controls

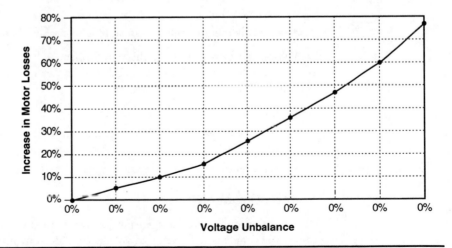

Figure 10–2   Effects of voltage unbalance on motor losses; Source: U.S. Department of Energy.

The following problems can be isolated to a particular circuit and may require a load-by-load survey to detect:

- Unequal impedance's in power-supply conductors, capacitors, or distribution wiring
- Certain kinds of motor defects

Constant loads can be checked by measuring voltage-to-ground on each phase with a hand-held voltmeter. Highly variable loads may require simultaneous measurement of all three phases, with monitoring over time. Monitoring instruments can periodically measure and record the voltage, current, power factor, and total harmonic distortion on each phase.

Proper system balancing can be maintained by:

- Checking that unloaded transformer voltage balance does not exceed a minimum 1.25 percent step per tap
- Checking and verifying electrical system single-line diagrams to ensure that single-phase loads are evenly distributed
- Regularly monitoring voltages on all phases to verify that a minimal unbalance exists
- Installing ground fault indicators
- Conducting annual infrared thermographic inspections
- Installing sensitive phase-voltage monitors

Before making future changes to your distribution system, consider the effect on the resulting phase-to-phase balance.

Over/Undervoltage (Voltage Deviation)

Proper system balancing can be maximized through the following steps:

- Check that unloaded transformer voltage balance does not exceed minimum steps (this is usually 1.25 or 2.5 percent step per tap)
- Check the electrical system single-line diagrams to verify that single-phase loads are evenly distributed.
- Regularly monitor voltages on all phases to verify that a minimal unbalance exists.

Voltage can be measured at the transformers and power panels to detect large-scale problems. Voltage measurements at the MCCs can be used to detect problems in the branch circuits and individual loads.

- Install required ground fault indicators.
- Conduct annual infrared thermographic inspections and eliminate high resistance connections.
- Install sensitive phase-voltage monitors.

Voltage Deviation

Three-phase induction motors generally operate most effectively at their rated voltage. The motor will compensate for reduced voltage conditions by increasing the current to match the torque requirements of the load. The higher currents result in higher I^2R losses within

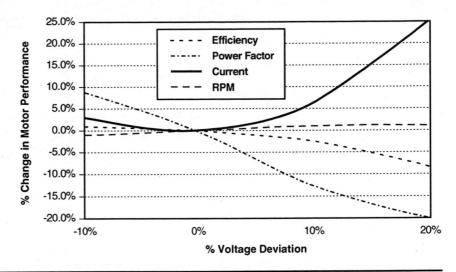

Figure 10–3   Voltage deviation effect on standard motor. Full performance at full load; Source: U.S. Department of Energy.

the windings and rotor. Low voltages can also prevent a motor from developing adequate staring torque. On the other hand, as voltage is increased, the magnetic current increases exponentially. Extreme overvoltage can increase core saturation and result in overheating. Starting current, starting torque, and breakdown torque all significantly increase with overvoltage conditions.

NEMA Standard MG I recommends that the voltage deviation for any motor not exceed +/− 10 percent. Although undervoltage raises the efficiency and power factor of standard motors, it also increases slip and temperature rise. Overvoltage reduces standard motor efficiency, power factor, and slip while increasing temperature rise. Energy-efficient motors tend to operate with increased efficiency and reduced temperature rise when supplied with overvoltage. Although motor performance depends on the particular motor design, Figures 10–3 and 10–4 graphically demonstrate how voltage deviation generally affects full load motor performance for a standard motor and energy-efficient motor, respectively. The trends shown in Figure 10–4 perhaps represent the upper limit for an energy efficient motor, rather than being typical. Over- or undervoltage conditions can result from:

- Incorrect selection of motors for the rated voltage. Examples are a 440-volt motor a 460-target-volt circuit, a 240-volt motor on a 208-volt circuit, or a 230/208-volt motor for a 208-circuit. A 230/208-volt-rated motor is actually a 230-volt motor.

- Incorrect transformer tap settings.

- Unequal branch line losses resulting in dissimilar voltage drops within the system. Often a panel will be supplied with a slight overvoltage in the hope of supplying the correct voltage to the MCCs. However, voltage drop differences can result in an overvoltage in some MCCs while others are undervoltage.

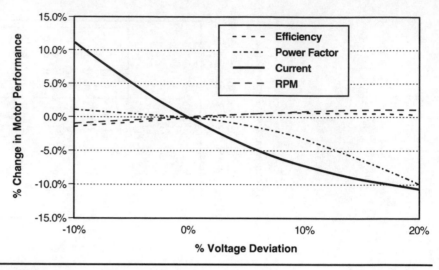

Figure 10–4 Voltage deviation effect on energy efficient motor performance at full load; Source: U.S. Department of Energy.

System voltage can be modified by:

- Adjusting the transformer tap settings
- Installing automatic tap-changing equipment where system loads vary greatly through the course of the day
- Installing power factor correction capacitors that raise the system voltage while correcting for power factor

Undersized Conductors

As plants expand, conductors sized for the original load are often undersized for the new loads they are required to carry. Undersized conductors present an additional resistive load on the circuit, similar to a poor connection. The cost of replacing or supplementing these conductors is often prohibitive from the standpoint of energy cost savings. However, the cost may be substantially less when done during the construction of expansion or retrofit projects.

Insulation Leakage

Conductor insulation tends to degrade over time. Degradation can be caused by:

- Extreme temperatures
- Mechanical abrasion
- Moisture contamination
- Chemical contamination
- Inadequate insulation levels

When insulation leakage occurs, it can either leak to ground or to another phase of a polyphase system. From the standpoint of energy loss, insulation leakage is analogous to a poor connection with the addition of the I^2R load imposed on the distribution system by the leakage current.

Insulation resistance varies with temperature and surface moisture even without any degradation. As a general rule, insulation resistance decreases by a factor of two for every 10° C temperature rise.

Insulation leakage can be tested with a megohmmeter where a DC voltage is applied to an open winding.

Harmonics

The increased use of solid-state equipment has given considerable rise to the concern over harmonics in power distribution systems. Harmonics were not significant at the six plants studied and, in general, are not thought to be a major contributor to losses in power distribution systems. However, harmonics do result in some energy loss.

Traditional loads, referred to as linear, operate with the standard sinusoidal voltage and current waveforms. However, a growing class of nonlinear loads draw a nonsinusoidal current. The distorted load current can then result in a distorted bus voltage, throughout the system, if the source impedance is high enough.

Typical loads that cause harmonic distortion are:

- Computers, especially PCs
- Computer terminals and work stations
- Computer peripherals and modems
- Word processors
- Copy machines
- Facsimile machines
- Teletype machines
- Telephone PBX
- Uninterruptable power supplies (UPSs)
- Adjustable frequency drives (AFDs)
- Rectifiers
- High-intensity discharge lighting
- Arc furnaces

The nonsinusoidal waveforms are actually a compilation of the fundamental sine wave and integral multiples of the fundamental. Total harmonic distortion is the ratio of the magnitude of all the combined harmonics to the magnitude of the fundamental.

Current harmonics create excess heat and voltage drop in the distribution system. Larger conductors must often be derated because higher frequency current tends to flow near the surface of the conductor, effectively increasing the resistance of the conductor. Other effects of current distortion include:

- Improper calibration of overload devices and meters
- Low power factor with possible utility penalties
- Reduced electrical system capacity
- Excess neutral current on three-phase, four-wire systems

- Overheating and failure of components such as transformers and circuits
- Overvoltage of system components such as AFDs and power factor capacitors
- Voltage distortion

The effects of voltage distortion are:

- Metering and relaying errors
- Unnecessary computer shutdowns
- Reduced power interruption tolerance
- Increased heating of motors, transformers, and switch gear
- Timing errors in electronic control and metering equipment associated with multiple zero crossings.

Harmonics can be blocked, diverted or canceled with the appropriate application of the following techniques:

- Delta–wye transformer.
- Isolating inductor
- Zig-zag autotransformer
- Harmonic trap filter
- Multiple trap filter
- Ferroresonant transformer
- Active filter

In the course of alleviating the effects of harmonics, these techniques surpress the harmonic currents at their source or convert them into heat energy.

Data Gathering

Constructing the Motor List

The electrician/engineer should methodically list all of the candidate motors in the plant. An information file should be created for each motor. Replacement alternatives should be examined and a contingent plan of action defined. Then when a motor failure occurs, the preferred repair or replacement plan can be put into effect in a timely fashion.

The electrician should divide the plant into logical areas and make a list of motors to be reviewed. Motors that are significant energy users should head the list. Motors that operate for extended periods of time and larger motors should also be on the list. Conversely, motors that run intermittently should be placed toward the end of the list.

Each plant will have to establish appropriate thresholds. Typical selection criteria include:

- Three-phase, NEMA Design B motors
- 10 to 600 HP
- 2,000 hours per year of operation or more
- Constant load (not intermittent, cyclic or fluctuating)
- Older and/or rewound standard efficiency motors
- Easy access
- A readable nameplate

The objective is to sort through all the motors and prioritize the list using size and annual length of operating time as the principal criteria for priority. Once a short list of motors has been made, individual data collection can take place.

Acquiring Motor Nameplate Data

The motor analysis requires that information from the motor nameplate be entered onto an appropriate data form. A typical motor nameplate, indicated in Figure 1–1 (first chapter) contains both descriptive and performance-based data, such as full-load efficiency, power factor, amperage, and operating speed. As illustrated in the next section, this information can be used to determine both the load imposed on the motor by its driven equipment and the motor efficiency at its load point.

Depending on the motor age and manufacturer practices, not all of the desired information appears on every motor nameplate. It is not unusual for power factor and efficiency to be missing. When data are not present, the electrician must find the required data elsewhere. The motor manufacturer is a logical source.

The motor age in years and its rewind history should also be recorded. The electrician/engineer must obtain this data from company records or, in many cases, from the recollection of people who have worked at the plant and can recall motor histories. Identify the coupling type and describe the motor load (device being driven). Also describe driven-equipment speed and the nature of the load being served by the motor. *Motor Nameplate and Field Test Data Form* is presented in appendix A.

Load-Time Profiles

Annual operating hours can be estimated by constructing an operating time profile. Such a profile, included in the "Motor Nameplate and Field Test Data Form", requires the user to provide input regarding motor use on various shifts during work days, normal weekends, and holidays.

The nature of the load being served by the motor is also important. Motors that are coupled to variable speed drives and that operate with low load factors or that serve intermittent, cyclic, or randomly acting loads are not good candidates for cost-effective replacement with energy-efficient units.

Measuring Operating Values

In a three-phase power system it is necessary to measure the following at each motor feeder:

- Phase-to-phase voltage between all three phases.
- Current values for all three phases
- Power factor in all three phrases

An electrician is expected to have the following: a voltmeter, a clamp-on ammeter, a power factor meter, and a tachometer.

When the motor operates at a constant load, only one set of measurements is necessary. When the motor operates at two or three distinct levels, measurements are required at each level as the current and power factor values vary with changes in load level. The electrician is responsible for determining the weighted-average motor load when the motor encounters multiple load levels.

A motor that drives a random-acting load presents a difficult measurement problem. The electrician should take a number of measurements and estimate the current and power factor that best represents the varying load. These values are used with the voltage to determine the typical power required by the load.

Motor and driven equipment speeds must be measured as closely as possible with a contact or strobe tachometer. Motor speed is important, because a replacement motor should duplicate the existing motor speed. When driving centrifugal loads (fans and pumps), the motor load is highly sensitive to operating speed. An energy-efficient motor usually operates at a slightly higher speed than does a standard motor. The higher speed may result in an increase in speed-sensitive loads, which can negate savings due to improved motor efficiency. A speed comparison is necessary to properly evaluate an energy-efficient motor conservation opportunity.

Field measurements should be taken for each motor on the prioritized lists with values for each motor entered on the appropriate "Motor Nameplate and Field Test Data Form".

Data Gathering Approaches

Figure 10–5 is a diagram of a typical three-phase power system serving a "delta" motor load configuration. To evaluate the motor operation, it is necessary to collect nameplate data and to use a multimeter and analog power factor meter to record the voltage, amperage, and power factor on each service phase or leg. Readings are taken on all three legs and averaged. Figure 10–6 indicates how measurements are taken with hand-held instruments. It is also useful to use a strobe tachometer to measure the speed of both the motor and the driven equipment.

Power supplied to the motor can be measured directly with a single instrument when a "direct reading" meter is available. The direct reading meter uses current transformers and voltage leads to reliably sense power. Direct reading instruments are more costly than typical multimeters. Until recently, it was uncommon for direct-reading meters to be found in industrial plants. This situation is improving with the availability of digital direct-reading power meters.

Utilization Voltage

Utilization voltage should be checked first. A convenient place to take measurements is at a motor starter enclosure. The voltage unbalance should be calculated. The utilization voltage unbalance should not be greater than 1 percent. System voltage unbalance problems need to be corrected before valid motor analyses can take place.

Service Voltage

When the utilization voltage unbalance is greater than 1 percent the electrician must check the voltage values at the service entrance. Measurements should be made at or as close to the service entrance as possible. Service voltage unbalance greater than 1 percent should be brought to the attention of the local electric utility for correction. Data acquisition techniques discussed in this guidebook are intended for the secondary side of in-plant distribution system power transformers.

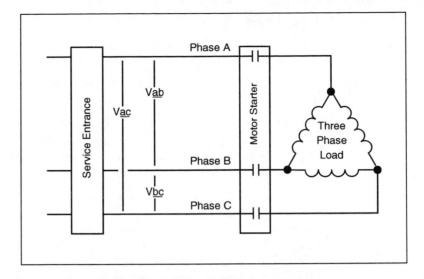

Figure 10–5   Industrial three phase circuit; Source: U.S. Department of Energy.

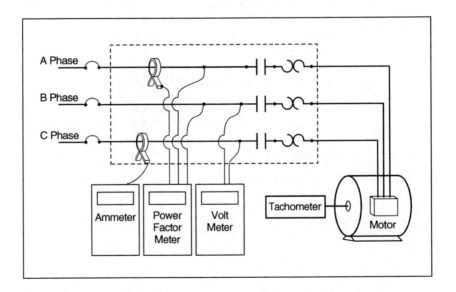

Figure 10–6   Instrument connection location; Source: U.S. Department of Energy.

Power Factor Measurements

Phase power factor is measured by clamping the current sensing element on one phase while attaching voltage leads to the other phases. Care must be taken to see that the proper voltage lead is used with the current sensing device.

Input Power

With measured parameters taken from hand-held instruments, the electrician can use the following equation to calculate the three-phase input power to the loaded motor.

$$W = (E \times I \times PF \times 1.73)/1{,}000$$

Where:
kW = input power
E = average value of line-to-line terminal voltage between the phases (Volts)
I = average value of the three line currents (mamperes)
PF = three-phase value (percent)/100

When the load is variable, the electrician must determine the average load imposed on the motor. That can be accomplished through long-term monitoring of the input power. If there are two load levels—for instance, a water supply pump motor that operates continuously but against two different static heads—both motor loads can be measured. The weighted average load is determined by timing the period when the motor encounters each load. When many load levels exist, the electrician must monitor loads over a period of time and develop an estimate for the weighted average load level.

When loads fluctuate randomly, hand-held instruments provide only a glimpse of the overall load profile. Random acting loads are difficult to assess with hand-held instruments. In random load situations, the use of recording meters with integrating capabilities is necessary to obtain valid data. Examples of various load types are given in *Table E* in the appendix.

Example Input Power Calculation

An existing motor is identified as a 40-Hp, 1800-rpm unit with an open drip-proof enclosure. The motor is 12 years old and has not been rewound.

The electrician makes the following measurements:
Measured Values:

Va-b = 467 V
Vb-c = 473 V
Vc-a = 469 V

Ia = 36 amps
Ib = 38 amps
Ic = 37 amps

PFa = 0.75
PFb = 0.78
PFc = 0.76

$$V \text{ average} = (467 + 473 + 469) \text{ V}/3 = 469.7 \text{ V}$$

$$I \text{ average} = (36 + 38 + 37) \text{ amps}/3 = 37 \text{ amps}$$

$$\text{Power Factor system} = (75\% + 78\% + 76\%)/3 = 76.3\%$$

Equation 10–1 reveals:

$$\text{Input Power} = 469.7 \text{ V} \times 37 \text{ amps} \times 0.763 \times 1.73/1{,}000 = 22.97\text{kW or}$$
$$22.97/0.746 = 30.8\text{hp}. \tag{10–1}$$

When the example motor is operated for an hour, 22.97 kWh of energy are consumed. At an energy rate of $0.04 per kWh, the cost of electrical energy for operating the motor is $0.92 per hour (22.97 kW × $0.04/kWh).

Motor Load Estimation Techniques

Part-load is a term used to describe the load served by the motor as compared to the full-load capability of the motor. Motor part-loads may be estimated through using input power, amperage, or speed measurements. See Equation 10–2.

Power Draw Measurements

Part-load may be quantified by comparing the input power to the motor while operating under load to the power required when the motor operates at rated capacity.
The relationship is:

$$\text{Part-Load (\%)} = (\text{Power measured/Power at rated load}) \times 100 \qquad (10\text{–}2)$$

The measured input power is determined from a direct reading meter or derived from voltage, amperage and power factor measurements (recall Equation 10–1). The power at rated load value is the power draw of the motor while it is delivering a shaft output equal to its nameplate rating.

Knowledge of the motor's full-load efficiency (e) is required to determine power at rated load. See the following equation.

$$\text{Power at rated load (kW)} = (\text{Motor hp} \times 0.746)/e \qquad (10\text{–}3)$$

Line Current Measurements

The *current load estimation method* is recommended when only amperage measurements are available. The amperage draw of a motor varies approximately linearly with respect to load down to about 50 percent of full-load. (See Figure 10–7) Below the 50 percent load point, due to reactive magnetizing current requirements, power factor degrades, and the amperage curve become increasingly nonlinear. In the low load region, current measurements are no longer a useful indicator of load.

$$\text{Motor part-load (\%)} = \text{Amps measured/Amps full-load nameplate}$$
$$\times \, (\text{volts measured/Volts nameplate}) \times 100 \qquad (10\text{–}4)$$

The Slip Method

The *slip method* is recommended when only motor operating speed measurements are available. The synchronous speed of an induction motor depends on the frequency of the power supply and on the number of poles for which the motor is wound. The higher the frequency, the faster a motor runs. The more poles the motor has, the slower it runs.

The actual speed of the motor is less than its synchronous speed, with the difference between the synchronous and actual speed referred to as slip. The amount of slip present is proportional to the load imposed upon the motor by the driven equipment. The motor load can be estimated with slip measurements as follows:

$$\text{Motorpart} - \text{load(\%)} = \frac{\text{Slip} \times 100}{\text{RPM sync} - \text{RPM fullload (nameplate)}} \qquad (10\text{–}5)$$

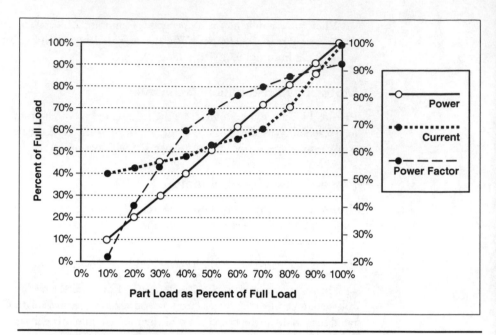

Figure 10–7 Relationship between power, current, power factor and motor load; Source: U.S. Department of Energy.

Where:

$$\text{Slip} = \text{RPM sync} - \text{RPM measured}$$

For Example:
Given:

$$\text{RPM sync} = 1{,}800 \quad \text{RPM measured} = 1{,}770$$

$$\text{RPM nameplate} = 1{,}750 \quad \text{Nameplate HP} = 25$$

Then:

$$\text{Slip} = 1{,}800 - 1{,}770 = 30$$

$$\text{Motor part-load} = 30/1800 - 1750 = 30 / 50 = .6$$

$$\text{Output hp} = .6 \times 25 \text{ hp} = 15 \text{ hp}$$

The speed/slip technique for determining motor part-load has been favored due to its simplicity and safety advantages. Most motors are constructed such that the shaft is accessible to a tachometer or a strobe light.

The accuracy of the slip method is, however, limited by multiple factors. The largest source of uncertainty related to use of the slip method is related to the 20 percent tolerance allowed by NEMA with respect to the manufacturers reporting of the nameplate full-load speed. Given this broad tolerance, manufacturers generally round their reported full-load speed values to some multiple of 5 rpm. Although 5 rpm is but a small percent of the full-load speed and may be thought of as insignificant, the slip method relies on the difference between full-load nameplate and synchronous speeds. Given a 40 rpm "correct" slip, a seemingly minor 5 rpm disparity causes a 12 percent change in calculated load. Slip also varies inversely with respect to the motor terminal voltage squared—and voltage is subject to a separate NEMA tolerance of + /− 10 percent at the motor terminals. A voltage correction factor can, of course, be inserted into the slip load equation. The revised slip load, assuming the motor is rated at 460 volts (nameplate), with a measured voltage of 482 volts is:

$$\text{Motor part load (\%)} = \text{Slip}/(\text{Rpm sync} - \text{RPM full load (nameplate)}) \times (460 \text{ volts}/482 \text{ volts})^2 \times 100 \qquad (10\text{–}6)$$

Advantages of using the current-based load estimation technique are that NEMA MG1-12.47 allows a tolerance of only 10 percent when reporting nameplate full-load current. In addition, motor terminal voltages only affect current to the first power, while slip varies with the square of the voltage.

Although the slip method is attractive for its simplicity, its precision should not be over-estimated. The slip method is generally not recommended for determining motor loads in the field.

Computerized Load and Efficiency Estimation Techniques

The Oak Ridge National Laboratory has developed MChEff, a computer program that uses an equivalent circuit method to estimate the load and efficiency of an in-service motor. Only nameplate data and a measurement of rotor speed is required to compute both the motor efficiency and load factor. Dynamometer tests have shown that the method produces efficiency estimates that average within +/− 3 percentage points of actual. This accuracy is valid for motor loads ranging from 25 to 100 percent of rated capacity. The program allows the user to enter optional measured data, such as stator resistance, to improve the accuracy of the efficiency estimate.

Motor Efficiency

The NEMA definition of full-load motor efficiency is:

$$\text{Motor Eff.(\%)} = 100 \times (746 \times \text{hp rated} \times \text{motor part load})/(\text{watts input})$$

Where:
   part load = % part load divided by 100
   HP (output) is the size rating of the motor
   watts (rated) is the value of input power measured at motor starter

By definition, a motor of a given rated horsepower is expected to deliver that quantity of power in a mechanical form at the motor shaft.

Figure 10–7 is a graphical depiction of the process wherein electrical energy is converted to mechanical energy. Motor losses are the difference between the input and output power. Once the motor efficiency has been determined and the input power is known, output power can be calculated.

NEMA design A, B, and E motors up to 500 HP in size are required to have a full-load efficiency value selected from a table of nominal efficiencies stamped on the nameplate. When a nameplate is missing or unreadable, it is up to the electrician to determine the efficiency value at the operating load point for an existing motor.

One way of obtaining efficiency information is to record significant nameplate data and then contact the motor manufacturer. With the style, type, and serial number, the manufacturer can identify approximately when the motor was manufactured. Often the manufacturer will have historical records and can supply nominal efficiency values as a function of load for a family of motors.

Efficiency determination is more difficult when the subject motor has been in service for a long time. It is not uncommon for the nameplate on the motor to be lost or painted over. When that is the case, it is almost impossible to locate efficiency information. Also, if the motor has been rewound, there is a high probability that the motor efficiency has changed.

Most analyses of motor energy conservation savings assume that the existing motor is operating at its nameplate efficiency. This assumption is reasonable above the 50 percent load point, because motor efficiencies generally peak at around three-fourths load with performance at 50 percent load almost identical to that at full load. Larger horsepower motors exhibit a relatively flat efficiency curve down to 25 percent of full load.

When an efficiency value is not stamped on a motor nameplate, motor efficiency estimates may be extracted from appendix B. Appendix B contains nominal efficiency values at full, 75, 50, and 25 percent load for typical standard-efficiency motors of various sizes and with synchronous speeds of 900, 1200, 1800, and 3600 rpm.

Using this efficiency and load estimation technique involves three steps. First, power, amperage or slip measurements are used to identify the load imposed on the operating motor. Then a motor part-load efficiency value, which is consistent with the approximated load, is obtained by interpolating from the data supplied in appendix B. A revised load estimate is then derived from both the power measurement at the motor terminals and the part-load efficiency value as follows:

$$\text{Load} = \frac{\text{kw measured} \times \text{Motor Efficiency at Load Point}}{\text{HP nameplate} \times 0.746} \qquad (10\text{--}7)$$

For rewound motors, an adjustment should be made to the efficiency values in Appendix B. Tests of rewound motors show the rewound motor efficiency to be less than that of the original motor. Generally, two points should be subtracted from your standard motor efficiency to reflect expected rewind losses on smaller motors (<40 HP) with one point subtracted for larger motors.

When to Buy Energy-Efficient Motors

Using readily available information such as motor nameplate capacity, operating hours, and electricity price you can quickly determine the simple payback that would result from selecting and operating an energy-efficient motor.

Using energy-efficient motors can reduce your operating costs in several ways. Not only does saving energy reduce your monthly electrical bill, it can postpone or eliminate the need to expand the electrical supply system capacity within your facility. On a larger scale, installing energy conserving devices allows your electrical utility to defer building expensive new generating plants, resulting in lower costs for you, the consumer.

Energy-efficient motors are higher-quality motors, with increased reliability and longer manufacturer's warranties, providing savings in reduced downtime, replacement, and maintenance costs.

There are three general opportunities for choosing energy-efficient motors: (1) when purchasing a new motor, (2) in place of rewinding failed motors, and (3) to retrofit an operable but inefficient motor for energy conservation savings.

Energy-efficient motors should be considered in the following instances:

- For all new installations
- When major modifications are made to existing facilities or processes
- For all new purchases of equipment packages that contain electric motors, such as air conditioners, compressors, and filtration systems
- When purchasing spares or replacing failed motors
- Instead of rewinding old, standard-efficiency motors
- To replace grossly oversized and underloaded motors
- As part of an energy management or preventative maintenance program
- When utility conservation programs, rebates, or incentives are offered that make energy-efficient motor retrofits cost-effective

Energy-Efficient Motor Performance and Price

The efficiency of a motor is the ratio of the mechanical power output to the electrical power input. This may be expressed as:

$$\text{Efficiency} = \frac{\text{Output}}{\text{Input}} = \frac{\text{Input} - \text{Losses}}{\text{Input}} = \frac{\text{Output}}{\text{Output} + \text{Input}}$$

In 1989, the National Electrical Manufacturers Association (NEMA) developed a standard definition for energy-efficient motors. The definition, designed to help users identify and compare electric motor efficiencies on an equal basis, includes a table of minimum nominal full-load efficiency values. A motor's performance must equal or exceed the nominal efficiency levels for it to be classified as "energy-efficient."

Motor Losses and Loss Reduction Techniques

The only way to improve motor efficiency is to reduce motor losses. Even though standard motors operate efficiently, with typical efficiencies ranging between 83 and 92 percent, energy-efficient motors perform significantly better. An efficiency gain from only 92 to 94 percent results in a 25 percent reduction in losses. Since motor losses result in heat rejected into the atmosphere, reducing losses can significantly reduce cooling loads on an industrial facility's air-conditioning system.

Motor energy losses can be segregated into five major areas, each of which is influenced by design and construction. One design consideration, for example, is the size of the air gap between the rotor and the stator. Large air gaps tend to maximize efficiency at the expense of power factor, while small air gaps slightly compromise efficiency while significantly improving power factor. Motor losses may be categorized as those that are fixed, occurring whenever the motor is energized, and remaining constant for a given voltage and speed, and those that are variable and increase with motor load.

These losses are described as follows:

1. Core loss represents energy required to magnetize the core material (hysteresis) and includes losses due to creation of eddy currents that flow in the core. Core losses are decreased through the use of improved permeability electromagnetic (silicon) steel and by lengthening the core to reduce magnetic flux densities. Eddy current losses are decreased by using thinner steel laminations.

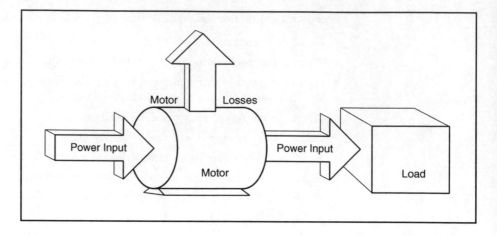

Figure 10–8 Depiction of motor losses; Source: U.S. Department of Energy.

2. Windage and friction losses occur due to bearing friction and air resistance. Improved bearing selection, air-flow, and fan design are employed to reduce these losses. In an energy-efficient motor, loss minimization results in reduced cooling requirements, so a smaller fan can be used. Both core losses and windage and friction losses are independent of motor load.

3. Stator losses appear as heating due to current flow (I) through the resistance (R) of the stator winding. This is commonly referred to as an I^2R loss. I^2R Losses can be decreased by modifying the stator slot design or by decreasing insulation thickness to increase the volume of wire in the stator.

4. Rotor losses appear as I^2R heating in the rotor winding. Rotor losses can be reduced by increasing the size of the conductive bars and end rings to produce a lower resistance, or by reducing the electrical current.

5. Stray load losses are the result of leakage fluxes induced by load currents. Both stray load losses and stator and rotor I^2R losses increase with motor load. Motor loss components are summarized in Figure 10–8. Loss distributions as a function of motor horsepower are given in Figure 10–9 while variations in losses due to motor loading are shown in Figure 10–10.

Definitions of Efficiencies

When evaluating motors on the basis of efficiency improvements or energy savings, it is essential that a uniform efficiency definition be used. It is often difficult to accurately compare manufacturers' published, quoted, or tested efficiencies, as various values are used in catalogs and vendor literature. Common definitions include:

- Average or nominal efficiency. These terms are identical and refer to the average full-load efficiency value obtained through testing a sample population of the same motor model. These are the most common standards used to compare motors.

- Guaranteed minimum or expected minimum efficiency. All motors purchased or a stated percentage of the motors purchased are guaranteed to have efficiencies that equal or exceed this full-load value. (Based on NEMA table.)

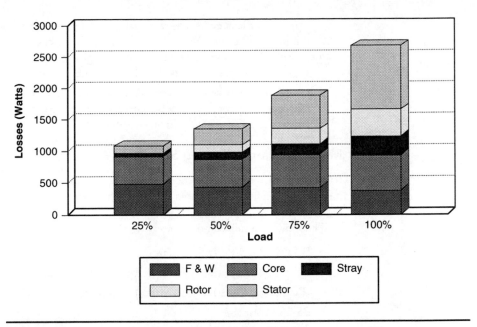

Figure 10–9   Motor losses versus load; Source: U.S. Department of Energy.

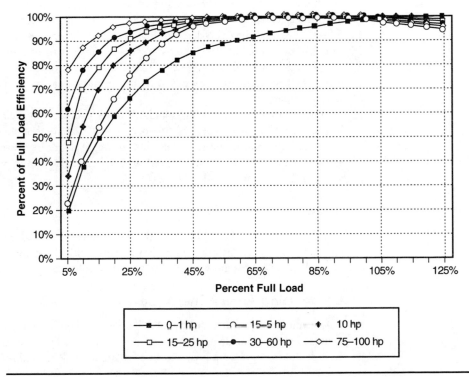

Figure 10–10   Motor part load efficiency as a function of percentage of full-load efficiency; Source: U.S. Department of Energy.

- Apparent efficiency. *Apparent efficiency* is the product of motor power factor and minimum efficiency. With this definition, energy consumption can vary considerably, as the power factor can be high while the efficiency is low. Specifications should not be based on "apparent" efficiency values.

- Calculated efficiency. This term refers to an average expected efficiency based on a relationship between design parameters and test results. Specifications should not be based on "calculated" efficiency values.

Motor Efficiency Testing Standards

It is critical that motor efficiency comparisons be made using a uniform product testing methodology. There is no single standard-efficiency testing method used throughout the industry.

The most common standards are:

- IEEE 112 -1984 (United States)
- EC 34-2 (International Electrotechnical Commission)
- JEC-37 (Japanese Electrotechnical Committee)
- BS-269 (British)
- C-39Q (Canadian standards Association)
- ANSI C50.20 same as IEEE. 112 (United States)

IEEE Standard 112-1984, Standard Test Procedure for Polyphase Induction Motors and Generators, is the common method for testing induction motors in the United States. Five methods for determining motor efficiency are recognized. The common practice for motors in the 1- to 125-HP size range is to measure the motor power output directly with a dynamometer, while the motor is operating under load. Motor efficiency is then determined by carefully measuring the electrical input and the mechanical power output.

The motor efficiency testing standards differ primarily in their treatment of stray load losses. The Canadian Standards Association (CSA) methodology and IEEE 112-Test Method B determine the stray load loss through an indirect process. The EC standard assumes stray load losses to be fixed at 0.5 percent of input, while the JEC standard assumes there are no stray load losses.

Determining Load Factor

To calculate the load factor, compare the power draw (obtained through wattmeter or voltage, amperage, and power factor measurements) with the nameplate rating of the motor. For a three-phase system, wattage draw = (PF × volts) (amps × 1,732).

Determining Annual Energy Savings

Before you can determine the annual dollar savings, you need to estimate the annual energy savings. For two similar motors operating at the same load, but having different efficiencies, the following equation is used to calculate the kW reduction.

$$\text{kw saved} = \text{hp} \times \text{L} \times 0.746 \times \frac{(100 - 100)}{(\text{Estd} - \text{EHE})} \tag{10-8}$$

Where:

HP = Motor nameplate rating

L = Load factor or percentage of full operating

Estd = Standard motor efficiency under actual load conditions

EHE = Energy-efficienct motor efficiency under actual load conditions

The kW savings are the demand savings. The annual energy savings are calculated as follows:

$$\text{kWh savings} = \text{kWsaved} \times \text{Annual Operating Hours} \qquad (10\text{--}9)$$

You can now use the demand savings and annual energy savings with utility rate schedule information to estimate your annual reduction in operating costs. Be sure to apply the appropriate seasonal and declining block energy charges. The total annual cost savings is equal to:

$$\text{Total savings} = \qquad (10\text{--}10)$$
$$(\text{kW saved} \times 12 \times \text{monthly demand charge}) + (\text{kWh savings} \times \text{energy charge})$$

Equations 10–9 and 10–10 apply to motors operating at a specified constant load. For varying loads, you can apply the energy savings equation to each portion of the cycle where the load is relatively constant for an appreciable period of time. The total energy savings is then the sum of the savings for each load period. Determine the demand savings at the peak load point. The equations are not applicable to motors operating with pulsating loads or for loads that cycle at rapidly repeating intervals.

Savings also depend on motor size and the gain in efficiency between a new high-efficiency motor and a new or existing unit. The performance gain for the energy-efficient motor is based on the difference between the average nominal full-load efficiencies for all energy-efficient motors on the market as compared to the average efficiency for all standard efficiency units.

Most industrial plant managers require that investments be recovered through energy savings within one to three years based on a simple payback analysis. The simple pay-back is defined as the period of time required for the savings from an investment to equal the initial or incremental cost of the investment. For initial motor purchases or the replacement of burned-out and unrewindable motors, the simple payback period for the extra investment associated with an energy-efficient motor purchase is the ratio of the price premium less any available utility rebate, to the value of the total annual electrical savings.

$$\text{simple payback years} = \frac{\text{Premium} - \text{utility rebate}}{\text{Total annual cost savings}} \qquad (10\text{--}11)$$

For replacements of operational motors, the simple payback is the ratio of the full cost of purchasing and installing a new energy-efficient motor relative to the total annual electrical savings.

$$\text{simple payback years} = \frac{\text{New motor cost} + \text{installation charge} - \text{utility rebate}}{\text{Total annual cost savings}} \qquad (10\text{--}12)$$

Example The following analysis for a 75-HP TEFC motor operating at 75 percent of full-rated load illustrates how to determine the cost-effectiveness of obtaining an energy-efficient versus a standard-efficiency motor savings for the initial purchase case.

Kilowatts saved:

$$\text{kW saved} = \text{hp} \times \text{load} \times 0 - 746 \times \frac{(100 - 100)}{(Estd - Ehe)}$$

$$= 75 \times .75 \times 0 - 746 \times \frac{(100 - 100)}{(91.6 - 94 - 1)}$$

$$= 1.21$$

Where:

EStd and *EHE* are the efficiencies of the standard motor and the alternative energy-efficient unit.

This is the amount of energy conserved by the energy-efficient motor during each hour of use. Annual energy savings are obtained by multiplying by the number of operating hours at the indicated load Energy saved:

$$\text{kWh savings} = \text{Hours of operation} \times \text{kw saved}$$

$$= 8,000 \text{ hours} \times 1.21$$

$$= 9,680 \text{ kwh/year}$$

Annual cost savings:

$$\text{Total cost savings} =$$

$$(\text{kWsaved} \times 12 \times \text{Monthly demand charge}) + (\text{kWhsavings} \times \text{Energy charge})$$

$$= 1.21 \times 12 \times \$5.35/\text{kW} + 9,680 \times \$0.03/\text{kWh}$$

$$= \$368$$

In this example, installing an energy-efficient motor reduces your utility billing by \$368 per year. The simple payback for the incremental cost associated with an energy-efficient motor purchase is the ratio of the discounted list price premium or incremental cost to the total annual cost savings. A list price discount of 75 percent is used in this analysis.

Recommendations for Motor Purchasers

As a motor purchaser you should be familiar with and use consistent sets of nomenclature. You should also refer to standard testing procedures.

- Insist that all guaranteed quotations are made on the same basis (i.e., nominal or guaranteed minimum efficiency.
- Prepare specifications that identify the test standard to be used to determine motor performance.
- Recognize the variance in manufacturing and testing accuracy and establish a tolerance range for acceptable performance.
- Comparison shop.
- Obtain an energy-efficient motor with a nominal efficiency within 1.5 percent of the maximum value available within an enclosure, speed, and size class.

Energy consumption and dollar savings estimates should be based on a comparison of nominal efficiencies as determined by IEEE 112—Method B for motors operating under appropriate loading conditions. Note that the NEMA marking standard only refers to efficiency values stamped on the motor nameplate. In contrast, manufacturers' catalogs contain values derived from dynamometer test data. When available, use catalog information to determine annual energy and dollar savings.

Choosing the Right Efficiency

Comparison shop when purchasing a motor, just as you would when buying other goods and services. Other things being equal, seek to maximize efficiency while minimizing the purchase price. Frequently, substantial efficiency gains can be obtained without paying a higher price.

Because so many motors exceed the minimum NEMA energy-efficiency standards, it is not enough to simply specify a "high efficiency" motor. Be certain to purchase a true "premium-efficiency" motor—an energy-efficient motor with the highest-available efficiency characteristics.

Obtaining Motor Efficiency Data

The slip, or difference between the synchronous and the operating speed for the motor, can be used to estimate the output kW, process load, and subsequently, the efficiency of the motor.

The synchronous speed of an induction motor depends on the frequency of the power supply and on the number of poles for which the motor is wound. The higher the frequency, the faster a motor runs. The more poles the motor has, the slower it runs. The synchronous speed (Ns) for a squirrel-cage induction motor is given by Equation 10–13.

$$Ns = \frac{60 \times 2f}{P} \qquad (10\text{–}13)$$

Where:
f = frequency of the power supply
p = poles for which the motor is wound

The actual speed of the motor is less than its synchronous speed. This difference between the synchronous and actual speed is referred to as slip. Slip is typically expressed as a percentage wire:

$$\text{Percent slip} = \frac{(\text{Synchronous speed} - \text{actual speed}) \times 100}{\text{Synchronous speed}}$$

You can now estimate motor load and efficiency with slip measurements.

$$\text{Slip} = \text{RPM sync} - \text{RPM measured}$$

$$\text{Motor load} = \frac{\text{slip}}{\text{RPM } \textit{sync} - \text{RPM } \textit{full load (nameplate)}}$$

$$\text{Approximate Output HP} = \text{Motor Load} \times \text{Nameplate HP} \qquad (10\text{–}14)$$

$$\text{Motor efficiency} = \frac{(0.746 \times \text{Output HP})}{\text{Measured input kW}}$$

An example:

Given:

$$\text{RPM synch} = 1800 \qquad \text{RPM measure} = 1770$$

$$\text{RPM nameplate} = 1750 \qquad \text{Nameplate hp} = 25$$

$$\text{Measured kW} = 13.1$$

Then:

$$\text{Slip} = 1800 - 1770 = 30$$

$$\text{Motor load} = \frac{30}{1800 - 1750} = \frac{30}{50} = .6$$

$$\text{Output hp} = 0.6 \times 25 = 15$$

$$\text{Motor efficiency} = \frac{(0.746 \times 15 \text{ hp}) \times 100\%}{13.1 \text{ kW}} = 85\%$$

Slip versus load curves can be obtained from your motor manufacturer. The slip technique for determining motor load and operating efficiency should not be used with rewound motors or with motors that are not operated at their design voltage.

It should be emphasized that the slip technique is limited in its accuracy. Although it cannot provide "exhaustive field efficiency testing results" it can be employed as a useful technique to identify and screen those motors that are oversized, underloaded, and operating at less than desired efficiencies.

Replacement of Operable Standard-Efficiency Motors

Based solely on energy savings, industrial users would typically find it not cost-effective to retrofit operable standard-efficiency motors with energy-efficient units. Such an action may, however, make sense if:

- Funding is available through a utility energy conservation program to partially offset the purchase price of the new energy-efficient motor.
- The standard-efficiency motor has been rewound several times.
- The standard-efficiency motor is oversized and underloaded.

Oversized and Underloaded Motors

When a motor has a significantly higher rating than the load it is driving, the motor operates at partial load. When this occurs, the efficiency of the motor is reduced. Motors are often selected that are grossly underloaded and oversized for a particular job.

Despite the fact that oversized motors reduce energy efficiency and increase operating costs, industries use oversized motors for the following:

- To ensure against motor failure in critical processes
- When plant personnel do not know the actual load and thus select a larger motor than necessary
- To build in capability to accommodate future increases in production
- To conservatively ensure that the unit has ample power to handle load fluctuations

- When maintenance staff replace a failed motor with the next larger unit if one of the correct size is not available
- When an oversized motor has been selected for equipment loads that have not materialized
- When process requirements have been reduced
- To operate under adverse conditions, such as voltage imbalance

As a general rule, motors that are undersized and overloaded have a reduced life expectancy with a greater probability of unanticipated downtime, resulting in loss of production. On the other hand, motors that are oversized—and thus, lightly loaded—suffer.

Maximum efficiency does not usually occur at fill load. As long as the motor is operating above 60 percent of rated load, the efficiency does not vary significantly. Motor efficiencies typically improve down to about 75 percent of full-rated load, then, especially for smaller motors, rapidly begin to decline below 40 percent of full load. It is almost always a good idea to downsize a motor that is less than 50 percent loaded.

Power factor declines sharply when the motor is operated below 75 percent of full-load amperage, especially in the smaller horsepower size ranges.

The cost penalties associated with using an oversized motor can be substantial and include:

- A higher motor purchase price
- Increased electrical supply equipment cost due to increased KVA and KVAR requirements
- Increased energy costs due to decreased part-load efficiency
- Power factor penalties

It is easiest to take measurements and properly size a motor driving a continuously running steady load. Be sure to take torque characteristics into consideration for intermittent or cyclic loading patterns. Also, be sure to provide adequate fan circulation and cooling for motors coupled to adjustable-speed loads or variable speed drives. Overheating is a particular concern at either reduced or full loads with the non ideal voltage and current waveforms encountered with electronic variable-frequency drives.

Electric load can be determined several ways. It can be directly measured with a power meter. A power factor meter and a clamp-on multimeter can be used in lieu of a power meter. Electric load can also be calculated from HP output. A stepwise procedure to determine HP output with only a tachometer is used on the data sheet. To determine electric load (i.e., input kW) the resulting HP output must be multiplied by 0.7457 and divided by the part-load efficiency. Part-load efficiency can be estimated from Figure 10–9, the manufacturer, or Motor Master Database.

For centrifugal loads, the replacement motor selected should be the next nameplate size above the motor output when operating under fully loaded conditions. It is recommended that voltage, amperage, kW draw, power factor, and slip be metered for a variety of motor operating conditions so the maximum load point can be known with confidence. The slip technique should not be used for rewound motors or motors operating at other than their design voltage.

The approach should not be used for motors driving conveyors or crushers where oversizing may be required to account for high startup torque, transient loads, or abnormal operating conditions. Most energy-efficient motors exhibit approximately the same locked rotor, breakdown, and rated-load torque characteristics as their standard-efficiency counterparts.

Speed, Design Voltage, Enclosure, Part-Load Efficiency, and Power Factor

Sensitivity of Efficiency Gains to Motor RPM

A motor's rotor must turn slower than the rotating magnetic field in the stator to induce an electrical current in the rotor conductor bars and thus produce torque. When the load on the motor increases, the rotor speed decreases. As the rotating magnetic field cuts the conductor bars at a higher rate, the current in the bars increases, which makes it possible for the motor to withstand the higher loading. Motors with Slip greater than 5 percent are specified for high inertia and high torque applications. NEMA Design B motors deliver a starting torque that is 150 percent of full-load or rated torque and run with a slip of 3 to 5 percent at rated load. Energy-efficient motors, however, are "stiffer" than equivalently sized standard motors and tend to operate at a slightly higher full-load speed.

On the average, energy-efficient motors rotate only 5 to 10 RPM faster than standard models. The speed range for available motors, however, exceeds 40 to 60 RPM. For centrifugal loads, even a minor change in the motor's full-load speed translates into a significant change in the magnitude of the load and energy consumption. The "Fan" or "Affinity Laws," indicated in Formula 10–1, show that the horsepower loading on a motor varies as the third power (cube) of its rotational speed. In contrast, the quantity of air delivered varies linearly with speed.

As summarized in Formula 10–2, a relatively minor 20-RPM increase in a motor's rotational speed, from 1740 to 1760 RPM, results in a 3.5 percent increase in the load placed upon the motor by the rotating equipment. A 40-RPM speed increase will increase air or fluid flow by only 2.3 percent, but can boost energy consumption by 7 percent, far exceeding any efficiency advantages expected from purchase of a higher efficiency motor. Predicted energy savings will not materialize—in fact, energy consumption will substantially increase.

This increase in energy consumption is especially troublesome when the additional air or liquid flow is not needed or useful. Be aware of the sensitivity of load and energy requirements to rated motor speed. Replacing a standard motor with an energy-efficient motor in a centrifugal pump or fan application can result in increased energy consumption if the energy-efficient motor operates at a higher RPM. A standard efficiency motor with a rated full-load speed of 1750 RPM should be replaced with a high-efficiency unit of like speed in order to capture the full energy conservation benefits associated with a high-efficiency motor retrofit. Alternatively, you can use sheaves or trim pump impellers so equipment operates at its design conditions.

$$\textbf{Fan Laws/Affinity Laws} \qquad \text{(Formula 10–1)}$$

$$\text{Law \#1} \quad \frac{\text{CFM 2}}{\text{CFM 1}} \quad \frac{\text{RPM 2}}{\text{RPM 1}}$$

Quantity (CFM) varies as fan speed (RPM).

$$\text{Law \# 2} \quad \frac{\text{P2}}{\text{P1}} \quad \frac{(\text{RPM 2})^2}{(\text{RPM 1}) \, 2^2}$$

Pressure (P) varies as the square of fan speed.

$$\text{Law \#3} \quad \frac{\text{HP 2}}{\text{HP 1}} \quad \frac{(\text{RPM 2})^3}{(\text{RPM 1})^3}$$

Horsepower (HP) varies as the cube of fan speed.

Sensitivity of Load to Motor RPM (Formula 10–2)

$$\frac{(1760)^3}{(1740)^3} = 3.5 \text{ percent horsepower increase}$$

$$\frac{(1780)^3}{(1740)^3} = 7.0 \text{ percent horsepower increase}$$

Operating Voltage Effects on Motor Performance

Generally, high-voltage motors have lower efficiencies than equivalent medium-voltage motors, because increased winding insulation is required for the higher voltage machines. This increase in insulation results in a proportional decrease in available space for copper in the motor slot. Consequently, I^2R losses increase.

Losses are also incurred when a motor designed to operate on a variety of voltage combinations (e. g., 208–230/460 volts) is operated with a reduced voltage power supply. Under this condition, the motor will exhibit a lower full-load efficiency, run hotter, slip more, produce less torque, and have a shorter life. Efficiency can be improved by simply switching to a higher voltage transformer tap.

If operation at 208 volts is required, an efficiency gain can be procured by installing an energy-efficient NEMA Design A motor. Efficiency, power factor, temperature rise, and slip are shown in Chart 25 for typical open-drip proof 10 HP, 1800 RPM Design B and Design A motors operated at both 230 and 208 volts.

Motor Speed and Enclosure Considerations

In general, higher-speed motors and motors with open enclosures tend to have slightly higher efficiencies than low-speed or totally enclosed fan-cooled units. In all cases, however, the energy-efficient motors offer significant efficiency improvements, and hence energy and dollar savings, when compared with the standard-efficiency models.

Efficiency Improvements at Part-Load Conditions

Energy-efficient motors perform better than their standard-efficiency counterparts at both full and partially loaded conditions. Efficiency improvements from use of a premium-efficiency motor actually increase slightly under half-loaded conditions. Although the overall energy conservation benefits are less for partially loaded versus fully loaded motors, the percentage of savings is relatively constant.

Power Factor Improvement

An induction motor requires both active and reactive power to operate. The active or true power, measured in kW, is consumed and produces work or heat. The reactive power, expressed in kVARs, is stored and discharged in the inductive or capacitive elements of the circuit. It establishes the magnetic field within the motor that causes it to rotate. The total power or apparent power is the product of the total voltage and total current in an AC circuit and is expressed in KVA. The total power is also the vector sum of the active and reactive power components. Power factor is the ratio of the active to the total power.

The electric utility must supply both active and reactive power loads. A low or "unsatisfactory" power factor is caused by the use of inductive (magnetic) devices and can indicate a possible low system electrical operating efficiency. Induction motors are generally the principal cause of low power factor because there are so many in use and they are not fully loaded.

When motors operate near their rated load, the power factor is high, but for lightly loaded motors the power factor drops significantly. This effect is partially offset, because the total current is less at reduced load. Thus, the lower power factor does not necessarily increase the peak KVA demand because of the reduction in load. Many utilities, however, levy a penalty or surcharge if a facility's power factor drops below 95 or 90 percent.

In addition to increased electrical billings, a low power factor may lower your plant's voltage, increase electrical distribution system line losses, and reduce the system's capacity to deliver electrical energy. Although motor full- and part-load power factor characteristics are important, they are not as significant as nominal efficiency. When selecting a motor, conventional wisdom is to purchase efficiency and correct for power factor. Low power factors can be corrected by installing external capacitors at the main plant service or at individual pieces of equipment. Power factor can also be improved and the cost of external correction reduced by minimizing operation of idling or lightly loaded motors and by avoiding operation of equipment above its rated voltage.

Power factors can usually be improved through replacement of standard with premium-efficiency motors. Power factors vary tremendously, however, based on motor design and load conditions. Although some energy-efficient motor models offer power factor improvements of 2 to 5 percent, others have lower power factors than typical equivalent standard motors. Even a high power factor motor affected significantly by variations in load A motor must be operated near its rated loading in order to realize the benefits of a high power factor design.

Motor Operation Under Abnormal Conditions

Motors must be properly selected according to known service conditions. Usual service conditions, defined in NEMA Standards Publication MG1-1987, *Motors and Generators,* include:

1. Exposure to an ambient temperature between 0° C and 40° C
2. Installation in areas or enclosures that do not seriously interfere with the ventilation of the machine
3. Operation within a tolerance of +/– 10 percent of rated voltage
4. Operation from a sine wave voltage source (not to exceed 10 percent deviation factor)
5. Operation within a tolerance of +5 percent of rated frequency
6. Operation with a voltage unbalance of 1 percent or less

Operation under unusual service conditions may result in efficiency losses and the consumption of additional energy. Both standard and energy-efficient motors can have their efficiency and useful life reduced by a poorly maintained electrical system. Monitoring voltage is important for maintaining high-efficiency operation and correcting potential problems before failures occur. Preventive maintenance personnel should periodically measure and log voltage at a motor's terminals while the machine is fully loaded.

Load Shedding

Energy and power savings can be obtained directly by shutting off idling motors to eliminate no-load losses. This action also greatly improves the overall system power factor, which, in turn, improves system efficiency. Typical no-load or idling power factors are in the 10 to 20 percent range. Load shedding is most effective for slower-speed (1800 RPM and less) motors used in low-inertia applications. Although it is possible to save energy by de-energizing the motor and restarting it when required, excessive starting, especially without soft-starting capacity, can cause overheating and increased motor failures.

Consideration must be given to thermal starting capability and the life expectancy of both motor and starting equipment. Motors 200 HP and less can only tolerate about 20 seconds of maximum acceleration time with each start. Motors should not exceed more than 150 start seconds per day. Starting limitations for motors greater than 200 HP should be obtained from the manufacturer.

Motor Selection Considerations

Overall motor performance is related to the following parameters:

Acceleration capabilities	Insulation class
Breakdown torque	Power factor
Efficiency	Service factor
Enclosure type	Sound level
Heating	Speed
Inrush current	Start torque

A good motor specification should define performance requirements and describe the environment within which the motor operates. As the purchaser, you should avoid writing design-based specifications that would require modification of standard components such as the frame bearing, design, rotor design, or insulation class.

Specification contents should include:

- Motor horsepower and service factors
- Temperature rise and insulation class
- Maximum starting current
- Minimum stall time
- Power factor range
- Efficiency requirement and test standard to be used
- Load inertia and expected number of starts

Environmental information should include:

- Abrasive or nonabrasive
- Altitude
- Ambient temperature
- Hazardous or nonhazardous
- Humidity level

You should specify special equipment requirements such as thermal protection, space heaters (to prevent moisture condensation), and whether standard or nonstandard conduit boxes are required.

GLOSSARY

ACROSS-THE-LINE. Method of motor starting that connects the motor directly to the supply line on starting or running; also called full voltage control.

AMBIENT TEMPERATURE. The temperature surrounding a device.

AMORTISSEUR WINDING. Consists of copper bars embedded in the cores of the poles of a synchronous motor. The copper bars of this special type of squirrel-cage winding are welded to end rings on each side of the rotor; used for starting only.

ARMATURE. A cylindrical, laminated iron structure mounted on a drive shaft; contains the armature winding.

ARMATURE WINDING. Wiring embedded in slots on the surface of the armature; voltage is induced in this winding on a generator.

ASA. American Standards Associations.

AUTOMATIC COMPENSATORS. Motor starters that have provisions for connecting three-phase motors automatically across 50 percent, 65 percent, 80 percent, and 100 percent of the rated line voltage for starting, in that order after preset timing.

AUTOTRANSFORMER. A transformer in which a part of the winding is common to both the primary and secondary circuits.

BRANCH CIRCUIT. The circuit conductors between the final overcurrent device protecting the circuit and the power outlet.

BREAKDOWN TORQUE (OF A MOTOR). The maximum torque that will develop with the rated voltage applied at the rated frequency, without an abrupt drop in speed. (ASA)

BRUSHLESS EXCITATION. The commutator of a conventional direct-connected exciter of a synchronous motor is replaced with a three-phase, bridge-type, solid-state rectifier.

GLOSSARY written with permission from Keljik, Delmar Publishing.

BUS. A conducting bar, of different current capacities, usually made of copper or aluminum.

BUSWAY. A system of enclosed power transmission that is current- and voltage-rated.

CAPACITOR-START MOTOR. A single-phase induction motor with a main winding arranged for direct connection to the power source and an auxiliary winding connected in series with a capacitor. The capacitor phase is in the circuit only during starting.

CENTRIFUGAL SWITCH. On single-phase motors, when the rotor is at normal speed, centrifugal force set up in the switch mechanism causes the collar to move and allows switch contacts to open, removing starting winding.

CIRCUIT BREAKER. A device designed to open and close a circuit by nonautomatic means and to open the circuit automatically on a predetermined overcurrent without injury to itself when properly applied within its rating.

COMMUTATOR. Consists of a series of copper segments that are insulated from one another and the mounting shaft; used on DC motors and generators.

CONTROLLER. A device or group of devices that governs, in a predetermined manner, the delivery of electric power to apparatus connected to it.

DC EXCITER BUS. A bus from which other alternators receive their excitation power.

DEFINITE-PURPOSE MOTOR. Any motor designed, listed, and offered in standard ratings with standard operating characteristics or mechanical construction for use under service conditions other than usual, or for use on a particular type of application.

DISCONNECTING MEANS (DISCONNECT). A device, or group of devices, or other means whereby the conductors of a circuit can be disconnected from their source of supply.

DISCONNECTING SWITCH. A switch that is intended to open a circuit only after the load has been thrown off by some other means. It is not intended to be opened under load.

DUAL VOLTAGE MOTORS. Motors designed to operate on two different voltage ratings.

DUTY CYCLE. The period of time in which a motor can safely operate under a load. *Continuous* means that the motor can operate fully loaded 24 hours a day.

DYNAMIC BRAKING. Using a DC motor as a generator, taking it off the supply line and applying an energy-dissipating resistor to the armature. Dynamic braking for an AC motor is accomplished by disconnecting the motor from the line and connecting DC power to the stator winding.

EFFICIENCY. The efficiency of all machinery is the ratio of the output to the input. Efficiency = output/input.

ELECTRIC CONTROLLER. A device, or group of devices, that governs, in some predetermined manner, the electric power delivered to the apparatus to which it is connected.

EUTECTIC ALLOY Metal with low and definite melting point; used in thermal overload relays; converts from a solid to a liquid state at a specific temperature; commonly called solder pot.

FEEDER. The circuit conductor between the service equipment, or the generator switchboard of an isolated plant, and the branch circuit overcurrent device.

FEELER GAUGE. A precision instrument with blades in thicknesses of thousandths of an inch for measuring clearances.

FULL-LOAD TORQUE (OF A MOTOR). The torque necessary to produce the rated horsepower of a motor at full-load speed.

EXCITER. A DC. generator that supplies the magnetic field for an alternator.

FIELD DISCHARGE SWITCH. Used in the excitation circuit of an alternator. Controls (through a resistor) the high inductive voltage created in the field coils by the collapsing magnetic field.

FLUX. Magnetic field; magnetism.

FREQUENCY. Cycles per second or hertz.

FUSE. An overcurrent protective device with a circuit-opening fusible part that is heated and severed by the passage of overcurrent through it.

GEAR MOTOR. A self-contained drive made up of a ball-bearing motor and a speed-reducing gear box.

GENERAL-PURPOSE MOTOR. Any open motor having a continuous 40C rating and designed, listed, and offered in standard ratings with standard operating characteristics and mechanical construction for use under usual service conditions

IEC. International Electrotechnical Commission, a European standards association that publishes standards for electrical equipment; used throughout the world.

GROWLER. An instrument consisting of an electromagnetic yoke and winding excited from an AC source; used to locate short-circuited motor coils.

INVERSE TIME. A qualifying term indicating that a delayed action is introduced purposely. This delay decreases as the operating force increases.

LOCKED ROTOR CURRENT (OF A MOTOR). The steady-state current taken from the line with the rotor locked (stopped) and with the rated voltage and frequency applied to the motor.

LOCKED ROTOR TORQUE (OF A MOTOR). The minimum torque that a motor will develop at rest for all angular positions of the rotor with the rated voltage applied at a rated frequency. (ASA)

MEGOHMMETER (MEGGER). An electrical instrument used to measure insulation resistance.

MEGOHMS. A unit of resistance equal to 1,000,000 ohms.

MOTOR CIRCUIT SWITCH (EXTERNALLY OPERATED DISCONNECT SWITCH, EXO). Motor branch circuit switch, rated in horsepower. Usually contains motor starting protection; safety switch.

MOTOR CONTROLLER. A device used to control the operation of a motor.

MOTOR MASTER DATABASE. Software for motor applications developed by the Motor Challenge division of the Department of Energy.

NEC. National Electrical Code.

NEMA. National Electrical Manufacturers Association

NONSALIENT ROTOR. A rotor that has a smooth cylindrical surface. The field poles (usually two or four) do not protrude above this smooth surface.

NORMAL FIELD EXCITATION. The value of DC field excitation required to achieve unity power factor in a synchronous motor.

OVERLOAD (HEATERS). Operation of equipment in excess of normal, full-load rating, or of a conductor in excess of rated ampacity, which, when it persists for a sufficient length of time, would cause damage or dangerous overheating.

OVERLOAD PROTECTION (RUNNING PROTECTION). Overload protection is the result of a device that operates on excessive current, but not necessarily on a short circuit, to cause the interruption of current flow to the device governed. Usually consists of thermal overload relay units inserted in series with the conductors supplying the motor.

POLARITY. The characteristic of a device that exhibits opposite quantities, such as positive and negative, within itself.

POLE. The north or south magnetic end of a magnet; a terminal of a switch; one set of contacts for one circuit of main power.

PULL-UP TORQUE (OF AC MOTOR). The minimum torque developed by the motor during the period of acceleration from rest to the speed at which break-down occurs. (ASA)

POWER FACTOR. The ratio of true power to apparent power. A power factor of 100 percent is the best electrical system.

RATING. The rating of a switch or circuit breaker includes: the maximum current, voltage of the circuit on which it is intended to operate, the normal frequency of the current, and the interrupting tolerance of the device.

ROTOR. The revolving part of an AC motor or alternator.

SALIENT FIELD ROTOR. Found on three-phase alternators and synchronous motors; field poles protrude from the rotor support structure. The structure is of steel construction and commonly consists of a hub, spokes, and a rim. This support structure is called a spider.

SELSYN. Abbreviation of the words self-synchronous. Selsyn units are special AC motors used primarily in applications requiring remote control. These units are also referred to as *synchros.*

SERVICE FACTOR. An allowable motor overload; the amount of allowable overload is indicated by a multiplier, which, when applied to a normal horsepower rating, dictates the permissible loading.

SLIP. In an induction motor, slip is the difference between the synchronous speed and the rotor speed, usually expressed as a percentage.

SLIP RINGS. Copper or brass rings mounted on, and insulated from, the shaft of an alternator or wound-rotor induction motor; used to complete connections between a stationary circuit and a revolving circuit.

STATOR. The stationary part of a motor or alternator; the part of the machine that is secured to the frame.

SYNCHRONOUS CAPACITOR. A synchronous motor operating only to correct the power factor and not driving any mechanical load.

SYNCHRONOUS SPEED. The speed at which the electromagnetic field revolves around the stator of an induction motor. The synchronous speed is determined by the frequency (hertz) of the supply voltage and the number of poles on the motor stator.

TORQUE. The rotating force of a motor shaft produced by the interaction of the magnetic fields of the armature and the field poles.

WOUND-ROTOR INDUCTION MOTOR. An AC motor consisting of a stator core with a three-phase winding, a wound rotor with slip rings, brushes and brush holders, and two end shields to house the bearings that support the rotor shaft.

RESOURCE CONTACTS

The following organizations and companies are valuable resources for your motor operation, design, and troubleshooting needs.

- **Motor Manufacture.** Siemens is a leading company with a reliable technical support group. Siemens also offers training courses in the application of motors. The dedicated staff at Siemens are extremely helpful in the area of motor applications, troubleshooting, and design.

  Siemens Energy and Automation
  100 Technology Dr.
  Alpharetta, GA 30202
  Tele: 770-740-3060

- **Testing and Troubleshooting.** NETA is the leading association in setting test standards for motors. NETA also offers informative books and magazines that are valuable resources for those troubleshooting and maintaining electric motors. NETA's educational resources and member support are outstanding.

  International Electrical Testing Association (NETA)
  P.O. Box 687, 106 Stone St.
  Morrison, CO 80465
  Tele: 303-697-8441
  Fax: 303-697-8431
  E-Mail: 10335.3401@ compuserve or
  http://www.netplace.net/neta/neta.htm

- **Motor Repair Shops.** EASA is the major trade association for motor rewind and repair. It is an international organization of more than 2,600 companies that sell, service, and repair industrial electric motors and related electro-mechanical equipment. Most members serve as distributors for various manufacturers of such equipment. EASA member companies account for about one-half of the rewind and repair centers in the United States.

In addition, EASA publishes an annual membership directory. This directory, which lists EASA members (distribution and repair facilities) by state and city, is a valuable resource for locating value-added motor distributors and rewind and repair centers in a given area. For each member, a code indicates the capabilities and services offered by each company, including what type of motor can be rebuilt, mechanical servicing and load testing capabilities, electric/electronic control servicing ability, and type of new products sold at the facility.

Electrical Apparatus Service Association (EASA)
1331 Baur Blvd.
St. Louis, Mo 63132
Tele: 314-993-2220
Fax: 314-993-1269

• **Wiring Method and Motor Design Applications.** The Square D Company has a good support staff. Their quick reference material is a valuable tool for the field electrician or engineer.

Square D Company
Automation and Control Business
P.O. Bx 27446
Raleigh, N.C. 27611

• **Motor Control.** The Electricians' Technical Reference books volume *Motor Control* is the most exhaustive control publication available. This book offers quick reference solutions to your troubleshooting, design, and application needs. This book is a valuable resource tool for engineers and electricians engaging in plant and construction operations.

Delmar Publishers
3 Columbia Circle
P.O.Box 15015
Albany, N.Y.
Tele: 1-800-998-7498

• **Energy-Efficient Motors.** The United States Department of Energy, Motor Challenge Section is a partnership program between the U.S. Department of Energy and the nation's industries. The program is committed to increasing the use of energy-efficient, industrial electric motor systems and related technologies.

The program is wholly funded by the U.S. Department of Energy and is dedicated to helping industry increase its competitive edge, while conserving the nation's energy resources and enhancing environmental quality.

The Motor Challenge Information Clearinghouse is your one-stop resource for objective, reliable, and timely information on electric motor-driven systems.

The Motor Challenge Clearinghouse
BOX 45171
Olympia, WA 98504-3171
Tele: 800-862-2086
Fax: 360-586-8303
Motor Challenge website on the Internet at www.motor.doe.gov.

- **Seminars.** Motors and Motor Controls seminars by Integrity Electrical Services are available for everyone from novices to the advanced level. These seminars offer motor operation, application, and control from a practical application. Most seminars are held on location at plant sites, universities, community colleges, and technical schools. These seminars also relate to power quality problems associated with motor application, design, and troubleshooting.

  I.E.S.
  60 Evergreen Park
  Florence, AL 35633
  Tele: 256-718-3320

- **SOFTWARE.** Motor master software contains a database of motor price and performance data that includes horsepower, speed, enclosure type, price, and efficiency. This software also contains analysis that calculates energy savings and paybacks. This software is offered free to Motor Challenge partners.

  MOTOR MASTER
  Washington State Energy Office
  809 Legion Way S.E.
  Olympia, WA 98504
  Tele: 206-956-2215

- **Software.** Electrical Calculations for Motor Applications software calculates the overcurrent and overload protection, conductor sizes, and applications of motors.

  POWER QUALITY CALCULATION TOOL
  60 Evergreen Park
  Florence, AL 35633
  Tele: 256-718-3320

- **OVERCURRENT PROTECTION.** Solid state and mechanical information and training is available from:

  Allen-Bradley
  5211 Linbar Dr., Suite 502
  Nashville, TN. 37211-1021
  Tele: 615-834-8200
  OR
  1201 South Second Street, Milwaukee, WI 53204
  Tele: 414-382-2000 Fax: 414-382-4444

CHARTS

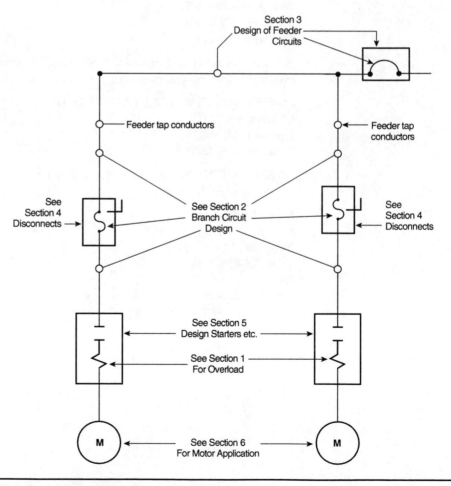

Chart 1a Motor design chart.

110v–250v
Single Phase Quick Reference Motor Data

HP #	Wire Size Copper 75° C		Full Load Current		Thermal Cir. Brk. Branch Circuit		Time Delay Fuse Branch Circuit		Min Size Starter	Conduit Size
	115v	230v	115v	230v	115v	230v	115v	230v		
1/6	14	14	4.4	2.2	15	15	7	3.5	00	1/2''
1/4	14	14	5.8	2.9	15	15	9	4.5	00	1/2''
1/3	14	14	7.2	3.6	15	15			00	1/2''
1/2	14	14	9.8	4.9	20	15	15	8	00	1/2''
3/4	14	14	13.8	6.9	25	15	20	10	00	1/2''
1	14	14	16	8	30	15	25	12	00	1/2''
1½	12	14	20	10	40	20	30	15	0	1/2''
2	10	14	24	12	50	25	30	17.5	0	1/2''
3	8	12	34	17	70	35	50	25	1	1/2''
5	4	10	56	28	90	60	80	40	2	3/4''
7½	3	8	80	40	110	80	100	60	2	3/4''
10		6		50		90		60	3	

Notes: All the above data is given to the minimum standard.

Chart 1b Single-phase quick reference motor data.

200–240 Volt
Three Phase Quick Reference Motor Design Data

HP	F.L.A.		Starter Size		Wire Size		Branch Circuit Inverse Time Circuit Breaker		Branch Circuit Time Delay Fuse		Conduit Size
	200v	230v	200v	230v	200v	230v	200v	230v	200v	230v	
1/2	2.3	2	00	00	14	14	15	15	3.5	3	1/2''
3/4	3.2	2.8	00	00	14	14	15	15	5	4	1/2''
1	4.1	3.6	00	00	14	14	15	15	6	5.6	1/2''
1½	6	5.2	00	00	14	14	15	15	9	8	1/2''
2	7.8	6.8	0	0	14	14	20	15	12	10	1/2''
3	11	9.6	0	0	14	14	30	20	17.5	15	1/2''
5	17.5	15.2	1	1	10	12	40	30	25	25	1/2''
7½	25.3	22	1	1	8	10	60	50	35	35	1/2''
10	32.2	28	2	2	8	8	70	70	45	45	1/2''
15	48.3	42	3	3	6	6	100	90	70	60	1''
20	62.1	54	3	3	4	4	125	100	90	80	1''
25	78.2	68	3	3	3	4	150	125	110	100	1''
30	92	80	4	3	2	3	175	150	125	125	1¼''
40	119.6	104	4	4	1/0	1	200	175	175	150	1½''
50	149.5	130	5	4	3/0	2/0	225	200	225	200	1½''
60	177.1	154	5	5	4/0	3/0	300	225	250	225	2''
75	220.8	192	5	5	300	250	400	350	300	300	2''
100	285.2	248	6	5	400	400	400	400	400	350	2½''
125	358.8	312	6	6	2–4/0	2–3/0	600	500	500	450	2–1½''
150	414	360	6	6	2–300	2–4/0	600	600	600	500	2–2''
200	552	480	7	6	2–500	2–350	800	800	600	600	2–2½''

Notes: All the above is given to the minimum. The designer may choose to exceed the above values.

Chart 1c Three-phase quick reference motor design data.

440v–600v
Three Phase Quick Reference Motor Design Data

HP	F.L.A. 460v	F.L.A. 575v	Starter Size 460v	Starter Size 575v	Wire Size 75° C C.U. 460v	Wire Size 75° C C.U. 575v	Branch Circuit Inv. Time C. Breaker 460v	Branch Circuit Inv. Time C. Breaker 575v	Branch Circuit Time Delay Fuse 460v	Branch Circuit Time Delay Fuse 575v	Conduit Size
$\frac{1}{2}$	1	.8	00	00	14	14	15	15	1.6	1.25	$\frac{1}{2}''$
$\frac{3}{4}$	1.4	1.1	00	00	14	14	15	15	2.25	1.8	$\frac{1}{2}''$
1	1.8	1.4	00	00	14	14	15	15	2.8	2.25	$\frac{1}{2}''$
$1\frac{1}{2}$	2.6	2.1	00	00	14	14	15	15	4	3.2	$\frac{1}{2}''$
2	3.4	2.7	00	00	14	14	15	15	5	4.5	$\frac{1}{2}''$
3	4.8	3.9	0	0	14	14	15	15	7	6.25	$\frac{1}{2}''$
5	7.6	6.1	0	0	14	14	15	15	12	10	$\frac{1}{2}''$
$7\frac{1}{2}$	11	9	1	1	14	14	30	20	17.5	15	$\frac{1}{2}''$
10	14	11	1	1	12	14	30	30	20	17.5	$\frac{1}{2}''$
15	21	17	2	2	10	10	40	35	35	25	$\frac{1}{2}''$
20	27	22	2	2	8	10	50	50	40	30	$\frac{1}{2}''$
25	34	27	2	2	8	8	90	60	50	40	$\frac{3}{4}''$
30	40	32	3	3	8	8	100	70	60	50	$\frac{3}{4}''$
40	52	41	3	3	6	6	125	90	80	60	$1''$
50	65	52	3	3	4	6	150	100	100	80	$1''$
60	77	62	4	4	3	4	150	100	125	90	$1\frac{1}{4}''$
75	96	77	4	4	1	3	200	125	150	100	$1\frac{1}{4}''$
100	124	99	4	4	2/0	1	250	175	200	150	$1\frac{1}{2}''$
125	156	125	5	5	3/0	2/0	300	200	225	175	$1\frac{1}{2}''$
150	180	144	5	5	4/0	3/0	300	225	250	200	$2''$
200	240	192	5	5	350	250	350	300	350	300	$2\frac{1}{2}''$

Notes: All the above data is given to the minimum standard.

Chart 1d   Three-phase quick reference motor design data.

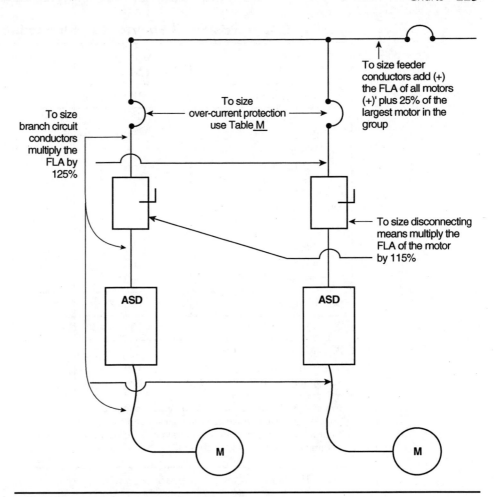

To size feeder conductors add (+) the FLA of all motors (+)' plus 25% of the largest motor in the group

To size over-current protection use Table M

To size branch circuit conductors multiply the FLA by 125%

To size disconnecting means multiply the FLA of the motor by 115%

ASD

ASD

M

M

Chart 1e   Adjustable speed drive.

Starting Characteristics of Squirrel Cage Induction Motors

Starting Method	Voltage At Motor	Line Current	Motor Torque
Full-Voltage Value	100	100	100
Autotransformer			
80% tap	80	64*	64
65% tap	65	42*	42
50% tap	50	25*	25
Primary Resistor Typical Rating	80	80	64
Primary Reactor			
80% tap	80	80	64
65% tap	65	65	42
50% tap	50	50	25
Series-Parallel	100	25	25
Wye-Delta	100	33	33
Part-Winding (1/2-1/2)			
2 to 12 Poles	100	70	50
14 and more Poles	100	50	50

Soft start is also available using solid-state controls.
Consult manufacturer for voltage, current and torque rating.
* Autotransformer magnetizing current not included. Magnetizing current is usually less than 25 percent of motor full-load current.

Chart 1f   Starting Characteristics of Squirrel Cage Induction Motors.

Standard Size Fuses and Circuit Breakers

Fuse Sizes	Fuse and Circuit Breaker Sizes		
1A	15A	100A	600A
3A	20A	125A	700A
6A	25A	150A	800A
10A	30A	175A	1000A
601A	35A	200A	1200A
	40A	225A	1600A
	45A	250A	2000A
	50A	300A	2500A
	60A	350A	3000A
	70A	400A	4000A
	80A	450A	5000A
	90A	500A	6000A

Chart 2 Standard size fuses and circuit breakers.

Duty-Cycle Service

Classification of Service	Percentages of Nameplate Current Rating			
	5-Minute Rated Motor	15-Minute Rated Motor	30 & 60 Minute Rated Motor	Continuous Rated Motor
Short-Time Duty Operating valves, raising or lowering rolls, etc.	110	120	150	. . .
Intermittent Duty Freight and passenger elevators, tool heads, pumps, drawbridges, turntables, etc. For arc welders, see Section 630-21	85	85	90	140
Periodic Duty Rolls, ore- and coal-handling machines, etc................	85	90	95	140
Varying duty	110	120	150	200

Any motor application shall be considered as continuous duty unless the nature of the apparatus it drives is such that the motor will not operate continuously with load under any condition of use.

Chart 3 Duty–cycle service; Reprinted with permission from NFPA 70–1996, the National Electrical Code®, Copyright © 1995, National Fire Protection Association, Quincy, MA 02269. This reprinted material is not the complete and official position of the National Fire Protection Association on the referenced subject, which is represented only by the standard in its entirety. National Electrical Code® and NEC® are registered trademarks of the National Fire Protection Association, Inc., Quincy, MA 02269.

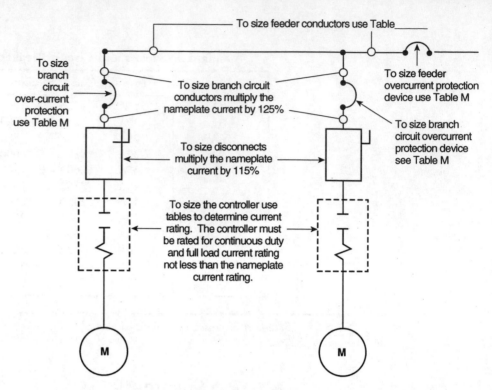

To size feeder conductors use Table_____

To size branch circuit over-current protection use Table M

To size branch circuit conductors multiply the nameplate current by 125%

To size feeder overcurrent protection device use Table M

To size branch circuit overcurrent protection device see Table M

To size disconnects multiply the nameplate current by 115%

To size the controller use tables to determine current rating. The controller must be rated for continuous duty and full load current rating not less than the nameplate current rating.

M   M

Note: For torque motors the rated current is the locked-rotor current. The name plate current must be used to determine the ampacity of the feeder and branch circuit conductors. The name plate current must be used to determine the ampere rating of overload, feeder and branch circuit protection.

Chart 4   Torque motors.

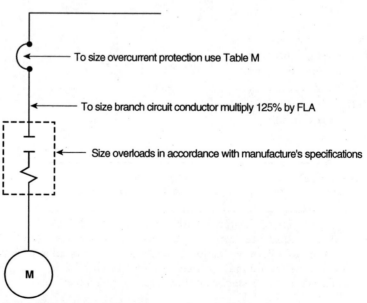

To size overcurrent protection use Table M

To size branch circuit conductor multiply 125% by FLA

Size overloads in accordance with manufacture's specifications

M

Part winding, start induction run or synchronous motor

Chart 5   Part winding motor.

Single Phase 115 Volt Motors		
Horse Power	**Ampere Rating**	**Switch Size Amperes**
1/20	1.13 to 1.33	
	1.34 to 1.44	
1/12	1.45 to 1.66	30
	1.67 to 1.80	Ampere
1/8	2.26 to 2.66	
	2.67 to 2.88	
1/6	2.89 to 3.33	
	3.34 to 3.60	
1/4	3.61 to 4.16	30
	4.17 to 4.50	Ampere
1/3	4.51 to 5.33	
	5.34 to 5.62	
1/2	5.63 to 6.66	
	6.67 to 7.20	
	7.21 to 8.00	
3/4	9.01 to 9.99	30
	10.0 to 10.8	Ampere
	10.9 to 12.5	
1	12.6 to 13.3	

Single Phase 230 Volt Motors		
Horse Power	**Ampere Rating**	**Switch Size Amperes**
1/3	2.26 to 2.66	
	2.67 to 2.88	
1/2	2.89 to 3.33	30
	3.34 to 3.60	Ampere
	3.61 to 4.16	
3/4	4.51 to 5.33	
	5.34 to 5.62	
1	5.63 to 6.66	
	6.67 to 7.20	
1 1/2	8.01 to 9.00	30
	9.01 to 9.99	Ampere
2	10.9 to 12.5	
	12.6 to 13.3	
3	15.7 to 18.0	30
	18.1 to 20.0	Ampere
5	26.1 to 31.2	60 Amp.

Chart 6 Size disconnects.

Three Phase 208 Volt Motors		
Horse Power	Ampere Rating	Switch
1/2	2.1	
3/4	3	
1	3.7	30
1 1/2	5.3	
2	6.9	
3	9.5	30
5	15.9	30
7 1/2	23.3	60
10	28.6	60
15	42.3	60
20	55	100
25	68	100
30	83	200
40	110	200
50	132	200
60	159	400
75	196	400
100	260	400
125	328	600
150	381	600

Three Phase 220 Volt Motors		
Horse Power	Ampere Rating	Switch Size Amperes
1/2	2	
3/4	2.8	
1	3.5	30
1 1/2	5	
2	6.5	
3	9	30
5	15	30
7 1/2	22	60
10	27	60
15	40	60
20	52	100
25	64	100
30	78	200
40	104	200
50	125	200
60	150	400
75	185	400
100	246	400
125	310	600
150	360	600
200	480	600

Chart 6 *Continued.*

Three Phase 440 Volt Motors			Three Phase 550 Volt Motors		
Horse Power	Ampere Rating	Switch Size Amperes	Horse Power	Ampere Rating	Switch Size Amperes
$1/2$	1		$1/2$	.8	
$3/4$	1.4		$3/4$	1.1	
1	1.8	30	1	1.4	30
$1^1/2$	2.5		$1^1/2$	2.0	
2	3.3		2	2.6	
3	4.5		3	4	
5	7.5		5	6	
$7^1/2$	11	30	$7^1/2$	9	30
10	14		10	11	
15	20		15	16	
20	26	60	20	21	60
25	32	60	25	26	60
30	39	60	30	31	60
40	52	100	40	41	60
50	63	100	50	50	100
60	75	200	60	60	100
75	93	200	75	74	200
100	123	200	100	98	200
125	155	400	125	124	200
150	180	400	150	144	400
200	240	400	200	192	400

Chart 6   *Continued.*

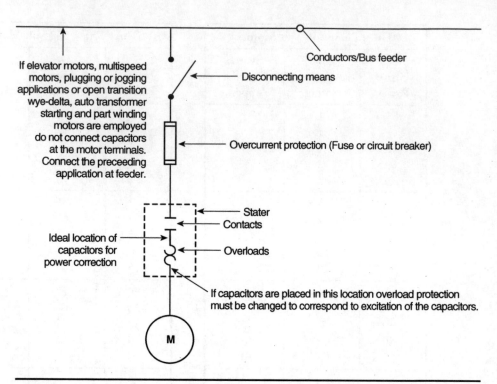

Chart 7 Capacitor placement for power factor correction.

Power-Factor Improvement
Capacitor Multipliers for Kilowatt Load

(To give capacitor KVAR required to improve power factor from original to desired value—see sample below.)

Original Power Factor, Percent	Desired Power Factor—Percent				
	100	95	90	85	80
60	1.333	1.004	0.849	0.713	0.583
62	1.266	0.937	0.782	0.646	0.516
64	1.201	0.872	0.717	0.581	0.451
66	1.138	0.809	0.654	0.518	0.388
68	1.078	0.749	0.594	0.458	0.328
70	1.020	0.691	0.536	0.400	0.270
72	0.964	0.635	0.480	0.344	0.214
74	0.909	0.580	0.425	0.289	0.159
76	0.855	0.526	0.371	0.235	0.105
77	0.829	0.500	0.345	0.209	0.079
78	0.802	0.473	0.318	0.182	0.052
79	0.776	0.447	0.292	0.156	0.026
80	0.750	0.421	0.266	0.130	
81	0.724	0.395	0.240	0.104	
82	0.698	0.369	0.214	0.078	
83	0.672	0.343	0.188	0.052	
84	0.646	0.317	0.162	0.206	
85	0.620	0.291	0.136		
86	0.593	0.264	0.109		
87	0.567	0.238	0.083		
88	0.540	0.211	0.056		
89	0.512	0.183	0.028		
90	0.484	0.155			
91	0.456	0.127			
92	0.426	0.097			
93	0.395	0.066			
94	0.363	0.034			
95	0.329				
96	0.292				
97	0.251				
99	0.143				

Assume total plant load is 100 KW qt 60 percent power factor. Capacitor KVAR rating necessary to improve power factor to 80 percent is found by multiplying KW (100) by multiplier in table (0.583), which gives KVAR (58.3). Nearest standard rating (60 KVAR) should be recommended.

Chart 8 Power factor improvement.

Single Voltage External Y-Connection

L1	L2	L3	Join
1	2	3	4 & 5 & 6

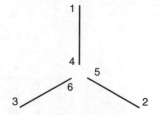

Single Voltage External Delta-Connection

L1	L2	L3
1, 6	2, 4	3, 5

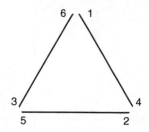

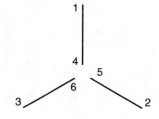

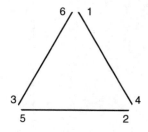

Single and Dual Voltage Star-Delta Connections

Single Voltage	Y-Connected Start Delta-Connected Run
Dual Voltage	Y-Delta-Connected (Voltage Ratio 1.723 to 1

Conn.		L1	L2	L3	Join
Y	Start or High Voltage	1	2	3	4 & 5 & 6
Delta	Run or Low Voltage	1, 6	2, 4	3, 5	——

Chart 9   Terminal markings and connections for single-speed, three-phase motors—6 leads.

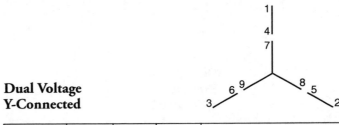

**Dual Voltage
Y-Connected**

Voltage	L1	L2	L3	Join
Low	1, 7	2, 8	3, 9	4 & 5 & 6
high	1	2	3	4 & 7, 5 & 8, 6 & 9

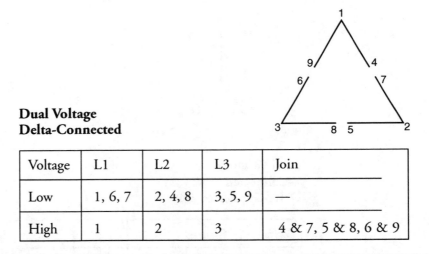

**Dual Voltage
Delta-Connected**

Voltage	L1	L2	L3	Join
Low	1, 6, 7	2, 4, 8	3, 5, 9	—
High	1	2	3	4 & 7, 5 & 8, 6 & 9

Chart 10   Terminal markings and connections for single-speed, three-phase motors—9 leads.

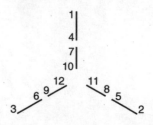

**Dual Voltage
External
Y-Connection**

Voltage	L1	L2	L3	Join
Low	1, 7	2, 8	3, 9	4 & 5 & 6, 10 & 11 & 12
High	1	2	3	4 & 7, 5 & 8, 6 & 9, 10 & 11 & 12

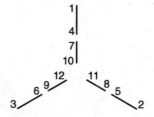

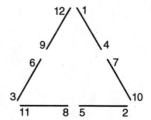

**Dual Voltage
Y-Connected Start
Delta-Connected Run**

Voltage	Conn.	L1	L2	L3	Join
Low	Start	1, 7	2, 8	3, 9	4 & 5 & 6, 10 & 11 & 12
	Run	1, 6, 7, 12	2, 4, 8,10	3, 5, 9, 11	—
High	Start	1	2	3	4 & 7, 5 & 8, 6 & 9, 10 & 11 & 12
	Run	1, 12	2, 10	3, 11	4 & 7, 5 & 8, 6 & 9

Chart 11 Terminal markings and connections for single-speed, three-phase motors—12 leads.

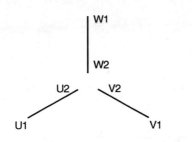 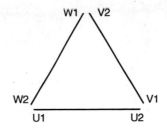

Single and Dual Voltage Star-Delta Connections

Single Voltage	Y-Connected Start Delta-Connected Run
Dual Voltage	Y-Delta-Connected (Voltage Ratio 1.723 to 1

Connection	L1	L2	L3	Together
Star	U1	V1	W1	U2 & V2 & W2
Delta	U1 & W2	V1 & U2	W1 & V2	None

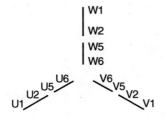 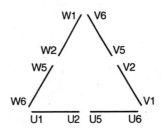

Dual Voltage Y-Connected Start Delta-Connected Run

Voltage	Connection	L1	L2	L3	Together
High	Star	U1	V1	W1	U2 & U5, V2 & V5, W2 & W5, U6 & V6 & W6
High	Delta	U1 & W6	V1 & U6	W1 & V6	U2 & U5, V2 & V5, W2 & W5
Low	Star	U1 & U5	V1 & V5	W1 & W5	U2 & V2 & W2 & U6, & V6 & W6
Low	Delta	U1 & U5 & W2 & W6	V1 & V5 & U2 & U6	W1 & W5 & V2 & V6	None

---

Chart 12   Terminal markings and connections for IEC single-speed, three-phase motors—6 and 12 lead.

Constant Torque Connection
Low-speed horsepower is half of high-speed Horsepower.*

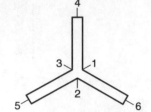

Speed	L1	L2	L3		Typical Connection
High	6	4	5	1 & 2 & 3 Join	2Y
Low	1	2	3	4-5-6 Open	1 Delta

Constant Horsepower Connection
Horsepower is the same at both speeds*

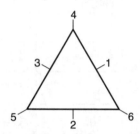

Speed	L1	L2	L3		Typical Connection
High	6	4	5	1-2-3 Open	1 Delta
Low	1	2	3	4 & 5 & 6 Join	2Y

Variable Torque Connection
Low-speed horsepower is one fourth of high-speed horsepower.*

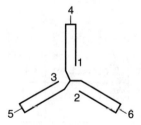

Speed	L1	L2	L3		Typical Connection
High	6	4	5	1 & 2 & 3 Join	2Y
Low	1	2	3	4-5-6 Open	1Y

*CAUTION: On European motors horsepower variance with speed may not be the same as shown above.

Chart 13 Terminal markings and connections for two-speed, three-phase motors, single winding with 6 leads.

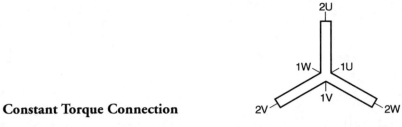

Constant Torque Connection

Speed	L1	L2	L3		Typical Connection
High	2W	2U	2V	1U-1V-1W Join	2Y
Low	1U	1V	1W	2U-2V-2W Open	1 Delta

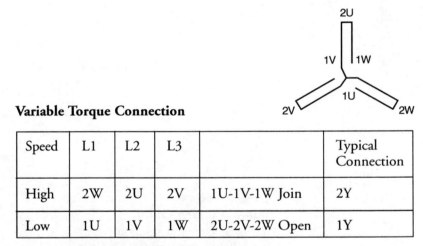

Variable Torque Connection

Speed	L1	L2	L3		Typical Connection
High	2W	2U	2V	1U-1V-1W Join	2Y
Low	1U	1V	1W	2U-2V-2W Open	1Y

IEC = International Electrotechnical Commission

Chart 14 Terminal markings and connections for IEC two-speed, three-phase motors, single winding with 6 leads.

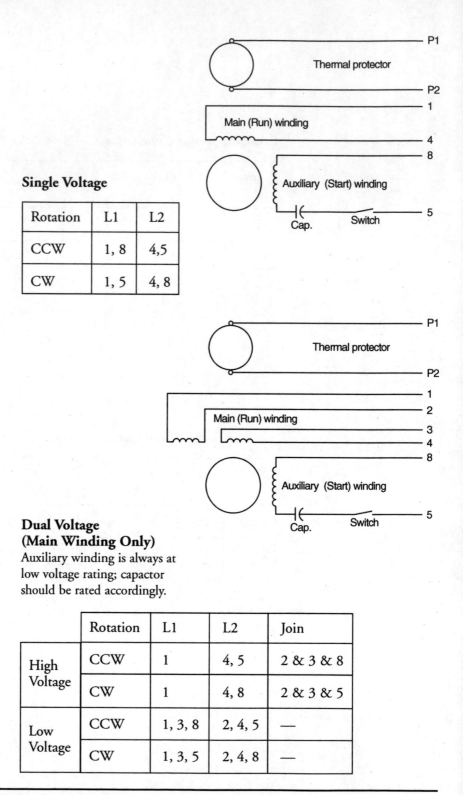

Single Voltage

Rotation	L1	L2
CCW	1, 8	4,5
CW	1, 5	4, 8

**Dual Voltage
(Main Winding Only)**
Auxiliary winding is always at
low voltage rating; capactor
should be rated accordingly.

	Rotation	L1	L2	Join
High Voltage	CCW	1	4, 5	2 & 3 & 8
	CW	1	4, 8	2 & 3 & 5
Low Voltage	CCW	1, 3, 8	2, 4, 5	—
	CW	1, 3, 5	2, 4, 8	—

Chart 15 Terminal markings and connections for capacitor-start, single phase-motors.

**Dual Voltage
(Main & Auxiliary
Winding)**
Capacitors in auxiliary
windings are rated for
lower voltage.

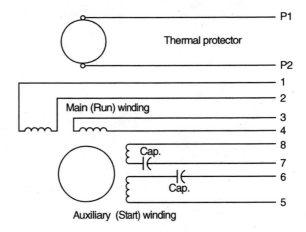

Main (Run) winding

Auxiliary (Start) winding

		L1	L2	Join
High Voltage	CCW	1, 8	4, 5	2 & 3, 6 & 7
	CW	1, 5	4, 8	2 & 3, 6 & 7
Low Voltage	CCW	1, 3, 6, 8	2, 4, 5, 7	—
	CW	1, 3, 5, 7	2, 4, 6, 8	—

The switch in the auxiliary winding circuit has been omitted from this
diagram. The connections to the switch must be made so that *both*
auxiliary windings become de-energized when the switch is open.

ROTATION: CCW—Counter-Clockwise
               CCW—Clockwise

The direction of shaft rotation can be determined by facing the end of
the motor opposite the drive.

Terminal Markings Identified by Color:
1-Blue   5-Black   P1-No color assigned
2-White   6-No color assigned P2-Brown
3-Orange   7-No color assigned
4-Yellow   8-Red

NEMA Standards MG1-2.41

Chart 16   Terminal markings and connections for capacitor-start, single-phase motors.

Shunt Generator

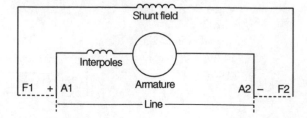

Compound Generator

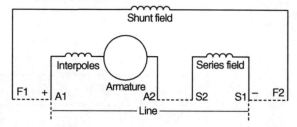

All connections are for counterclockwise rotation facing end opposite drive. For clockwise rotation, interchange A1 and A2.

Some manufacturers connect the interpole winding on the A2 side of armature.

For above generators, the shunt field may be either self-excited or separately excited. When self-excited, connections should be made as shown. When seperately excited, the shunt field is isolated from the other windings. When separately excited same polarities must be observed for given rotation.

Chart 17   Terminal markings and connections for D-C generators.

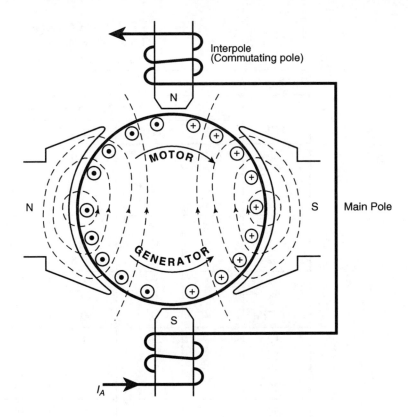

Diagram shows polarity of interpoles with respect to the polarity of the main poles.

For a MOTOR, the polarity of the interpole is the same as that of the main pole PRECEDING it in the direction of rotation.

For a GENERATOR, the polarity of the interpole is the same as that of the main pole FOLLOWING it in the direction of rotation.

Chart 18 Relationship of main and interpole polarities in D-C machines.

Full-Load Currents
Direct-Current Motors
(Running at Base Speed)

*For conductor sizing only.

HP	Armature Voltage Rating*					
	90V	120V	180V	240V	500V	550V
1/4	4.0	3.1	2.0	1.6		
1/3	5.2	4.1	2.6	2.0		
1/2	6.8	5.4	3.4	2.7		
3/4	9.6	7.6	4.8	3.8		
1	12.2	9.5	6.1	4.7		
1 1/2		13.2	8.3	6.6		
2		17	10.8	8.5		
3		25	16	12.2		
5		40	27	20		
7 1/2		58	39	29	13.6	12.2
10		76	51	38	18	16
15				55	27	24
20				72	34	31
25				89	43	38
30				106	51	46
40				140	67	61
50				173	83	75
60				206	99	90
75				255	123	111
100				341	164	148
125				425	205	185
150				506	246	222
200				675	330	294

Over 200 HP
Approx. Amps/HP   3.4   1.7   1.5

*These are average direct-current quantities. Branch-circuit conductors supplying a single motor shall have an ampacity not less than 125 percent of the motor full-load current rating.

Armature Current Varies Inversely as Applied Voltage.
Example: 40 HP Motor, 300 Volt Armature

$$\text{ARMATURE CURRENT} = 140 \times \frac{240}{300} = 112 \text{ AMPS}$$

Chart 19 Full-load currents. Reprinted with permission from NFPA 70–1996, the National Electrical Code®, Copyright © 1995, National Fire Protection Association, Quincy, MA 02269. This reprinted material is not the complete and official position of the National Fire Protection Association on the referenced subject, which is represented only by the standard in its entirety.

Minimum Full-Load Efficiencies
of Energy-Efficient Motors*
Open Motors

*As specified in the Energy Policy Act of 1992

HP	2 Pole Nominal Efficiency	4 Pole Nominal Efficiency	6 Pole Nominal Efficiency
1.0	—	82.5	80.0
1.5	82.5	84.0	84.0
2.0	84.0	84.0	85.5
3.0	84.0	86.5	86.5
5.0	85.5	87.5	87.5
7.5	87.5	88.5	88.5
10.0	88.5	89.5	90.2
15.0	89.5	91.0	90.2
20.0	90.2	91.0	91.0
25.0	91.0	91.7	91.7
30.0	91.0	92.4	92.4
40.0	91.7	93.0	93.0
50.0	92.4	93.0	93.0
60.0	93.0	93.6	93.6
75.0	93.0	94.1	93.6
100.0	93.0	94.1	94.1
125.0	93.6	94.5	94.1
150.0	93.6	95.0	94.5
200.0	94.5	95.0	94.5

(Compliance on non-exempted newly manufactured or imported general purpose motors to be fully implemented by 1997.)

Chart 20 Minimum full-load efficiencies of energy-efficient motors—open motors.

Minimum Full-Load Efficiencies
of Energy-Efficient Motors*
EnclosedMotors

*As specified in the Energy Policy Act of 1992

HP	2 Pole Nominal Efficiency	4 Pole Nominal Efficiency	6 Pole Nominal Efficiency
1.0	75.5	82.5	80.0
1.5	82.5	84.0	85.5
2.0	84.0	84.0	86.5
3.0	85.5	87.5	87.5
5.0	87.5	87.5	87.5
7.5	88.5	89.5	89.5
10.0	89.5	89.5	89.5
15.0	90.2	91.0	90.2
20.0	90.2	91.0	90.2
25.0	91.0	92.4	91.7
30.0	91.0	92.4	91.7
40.0	91.7	93.0	93.0
50.0	92.4	93.0	93.0
60.0	93.0	93.6	93.6
75.0	93.0	94.1	93.6
100.0	93.6	94.5	94.1
125.0	94.5	94.5	94.1
150.0	94.5	95.0	95.0
200.0	95.0	95.0	95.0

(Compliance on non-exempted newly manufactured or imported
general purpose motors to be fully implemented by 1997.
Explosion-proof UL-labeled motors need not comply until 1999.)

Chart 21   Minimum full-load efficiencies of energy-efficient motors—enclosed motors.

General Effect of Voltage and Frequency Variations on Induction Motor Characteristics

Characteristic	Voltage		Frequency	
	110%	90%	105%	95%
Starting torque	Up 21%	Down 19%	Down 10%	Up 11%
Maximum Torque	Up 21%	Down 19%	Down 10%	Up 11%
Percent Slip	Down 15–20%	Up 20–30%	Up 10–15%	Down 5–10%
Efficiency Full Load	Down 0–3%	Down 0–2%	Up Slightly	Down Slightly
¾ Load	0–Down Slightly	Little Change	Up Slightly	Down Slightly
½ Load	Down 0–5%	Up 0–1%	Up Slightly	Down Slightly
Power Factor Full Load	Down 5–15%	Up 1–7%	Up Slightly	Down Slightly
¾ Load	Down 5–15%	Up 2–7%	Up Slightly	Down Slightly
½ Load	Down 10–20%	Up 3–10%	Up Slightly	Down Slightly
Full-Load Current	Down Slightly to Up 5%	Up 5–10%	Down Slightly	Up Slightly
Starting Current	Up 10%	Down 10%	Down 5%	Up 5%
Full-Load Temperature Rise	Up 10%	Down 10–15%	Down Slightly	Up Slightly
Maximum Overload Capacity	Up 21%	Down 19%	Down Slightly	Up Slightly
Magnetic Noise	Up Slightly	Down Slightly	Down Slightly	Up Slightly

Chart 22 General effect of voltage and frequency variations on induction motor characteristics.

Comparison of Methods of Starting Squirrel Cage Induction Motors

Starter Type	% Full-Voltage Value		
	Voltage at Motor	Line Current	Motor Output Torque
Full Voltage	100	100	100
Autotransformer			
80 pc tap	80	64*	64
65 pc tap	65	42*	42
50 pc tap	50	25*	25
Primary-reactor			
80 pc tap	80	80	64
65 pc tap	65	65	42
50 pc tap	50	50	25
Primary-resistor Typical rating	80	80	64
Part-winding			
Low-speed motors (½–½)	100	50	50
High-speed motors (½–½)	100	70	50
High-speed motors (⅔–⅓)	100	65	42
Wye Start-Delta Run (⅓–⅓)	100	33	33

Soft start is also available using solid-state controls. Consult manufacturer for voltage, current and torque.

*Autotransformer magnetizing current not included. Magnetizing current usually less than 25 percent motor full-load current.

Chart 23 Comparison of methods of starting squirrel cage induction motors.

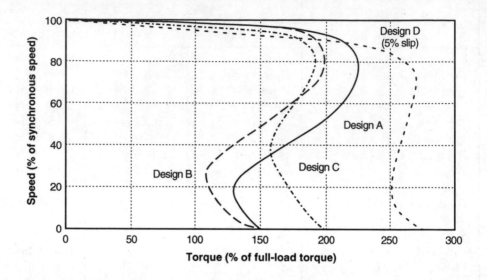

NEMA Design	Starting Current	Locked Rotor	Breakdown Torque	% Slip
B	Medium	Medium Torque	High	Max. 5%
	Applications: Normal starting torque for fans, blowers, rotary pumps, unloaded compressors, some conveyors, metal cutting machine tools, misc. machinery. Slight speed change with changing load.			
C	Medium	High Torque	Medium	Max. 5%
	Applications: High inertia starts, such as large, centrifugal blowers, fly wheels, and crusher drums. Loaded starts, such as piston pumps, compressors and conveyors. Slight speed change with changing load.			
D	Medium	Extra High Torque	Low	5% or more
	Applications: Very high inertia and loaded starts. Also, considerable variation in load speed. Punch presses, shears and forming machine tools. Cranes, hoists, elevators, and oil well pumping jacks.			

NEMA Design A is a variation of Design B having higher locked rotor current.

Chart 24   General speed-torque characteristics—three-phase induction motors.

When the line voltages applied to a polyphase induction motor are not equal, unbalanced currents in the stator windings will result. A small percentage voltage unbalance will result in a much larger percentage current unbalance. Consequently, the temperature rise of the motor operating at a particular load and percentage voltage unbalance will be greater than for the motor operating under the same conditions with balanced voltages.

Should voltages be unbalanced, the rated horsepower of the motor should be multiplied by the factor shown in the graph below to reduce the possibility of damage to the motor. Operation of the motor above a 5 percent voltage unbalance condition is not recommended.

Alternating-current, polyphase motors shall operate successfully under running conditions at rated load when the voltage unbalance at the motor terminals does not exceed 1 percent. Performance will not necessarily be the same as when the motor is operating with a balanced voltage at the motor terminals.

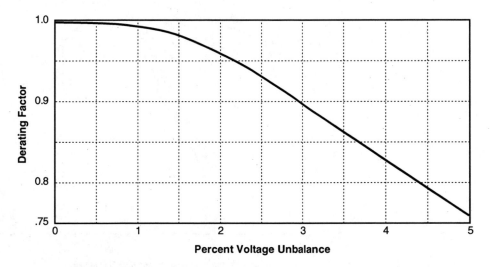

Percent Voltage Unbalance

Medium Horsepower Motors Derating Factor Dut to Unbalanced Voltage

Derating Factor = [100 − (Unbalanced %)^2]%

$$\frac{\text{Percent Voltage}}{\text{Unbalanced}} = 100 \times \frac{\text{Maximum Voltage Deviation from Average Voltage}}{\text{Average Voltage}}$$

Example—With voltages of 220, 214, and 210, the average is 215, the maximum deviation from the average is 5, and the

$$\frac{\text{Percent Voltage}}{\text{Unbalanced}} = 100 \times \frac{5}{215} = 2.3 \text{ percent}$$

Derating Factor = 100 − (2.3%)^2 = 94.7%

Reference NEMA Standards MG 1–12.45 & MG 1–14.35.

Chart 25   Effects of unbalanced voltage on motor performance.

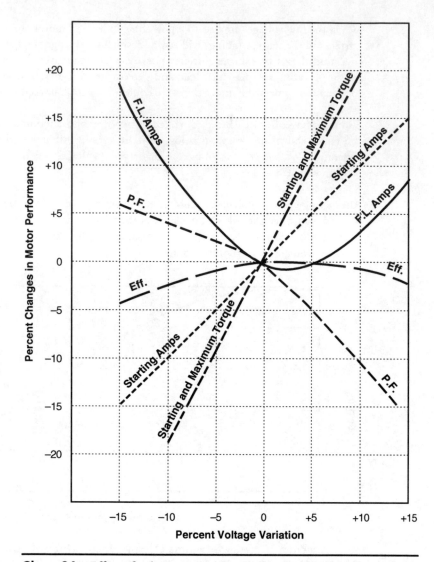

Chart 26 Effect of voltage variation on induction motor characteristics.

NEMA Size Starters for Single-Phase Motors

NEMA Size	Maximum Horsepower Single-Phase Motors Full-Voltage Starting	
	115 V.	230 V.
00	$1/3$	1
0	1	2
1	2	3
$1\frac{1}{2}$	3	5
2	—	$7\frac{1}{2}$
3	—	15

Chart 27 NEMA size starters for single-phase motors; Reprinted with permission from NFPA 70-1996, the National Electrical Code®, Copyright © 1995, National Fire Protection Association, Quincy, MA 02269. This reprinted material is not the complete and official position of the National Fire Protection Association on the referenced subject, which is represented only by the standard in its entirety.

Enclosures for Starters

Type	NEMA Enclosure
1	General Purpose—Indoor
2	Dripproof—Indoor
3	Dusttight, Raintight, Sleettight—Outdoor
3R	Rainproof, Sleet Resistant—Outdoor
3S	Dusttight, Raintight, Sleetproof—Outdoor
4	Watertight, Dusttight, Sleet Resistant—Indoor & Outdoor
4X	Watertight, Dusttight, Corrosion-Resistant—Indoor & Outdoor
5	Dusttight—Indoor
6	Submersible, Watertight, Dusttight, Sleet Resistant—Indoor & Outdoor
6P	Watertight—Prolonged Submersion—Indoor & Outdoor
7	Class I, Group A, B, C or D Hazardous Locations, Air-Break—Indoor
8	Class I, Group A, B, C or D Hazardous Locations, Oil-Immersed—Indoor
9	Class II, Group E, F or G Hazardous Locations, Air-Break—Indoor
10	Bureau of Mines
11	Corrosion-Resistant and Dripproof, Oil-Immersed—Indoor
12	Industrial Use, Dusttight and Driptight—Indoor
12K	Industrial Use, Dusttight and Driptight, with Knockouts—Indoor
13	Oiltight and Dusttight—Indoor

Reference NEMA Standard 250–1988.

Type	IEC Enclosure
2,3	General Purpose—Indoor
21,31	Dripproof—Indoor
34	Rainproof—Outdoor
65	Industrial Use—Indoor
66	Seawater Tight
68	Submersible

Chart 28 Enclosures for starters.

NEMA Size Starters for Three-Phase Motors

NEMA Size	Maximum Horsepower Polyphase Motors											
	Full Voltage Starting			Auto-Transformer Starting			Part-Winding Starting			Wye-Delta Starting		
	200V	230V	460V 575V	200V	230V	460V 575V	200V	230V	460V 575V	200V	230V	460V 575V
00	1½	1½	2	—	—	—	—	—	—	—	—	—
0	3	3	5	—	—	—	—	—	—	—	—	—
1	7½	7½	10	7½	7½	10	10	10	15	10	10	15
2	10	15	25	10	15	25	20	25	40	20	25	40
3	25	30	50	25	30	50	40	50	75	40	50	75
4	40	50	100	40	50	100	75	75	150	60	75	150
5	75	100	200	75	100	200	150	150	350	150	150	300
6	150	200	400	150	200	400	—	300	600	300	350	700
7	—	300	600	—	300	600	—	450	900	500	500	1000
8	—	450	900	—	450	900	—	700	1400	750	800	1500
9	—	800	1600	—	800	1600	—	1300	2600	1500	1500	3000

Chart 29   NEMA size starters for three-phase motors.

NEMA Code Letters for A-C Motors

NEMA Code Letter	Locked Rotor KVA Per HP*	NEMA Code Letter	Locked Rotor KVA Per HP*
A	0–3.15	L	9.0–10.0
B	3.15–3.55	M	10.0–11.2
C	3.55–4.0	N	11.2–12.5
D	4.0 –4.5	P	12.5–14.0
E	4.5 –5.0	R	14.0–16.0
F	5.0 –5.6	S	16.0–18.0
G	5.6 –6.3	T	18.0–20.0
H	6.3 –71.	U	20.0–22.4
J	7.1 –8.0	V	22.4 and up
K	8.0 –9.0		

NEMA Standards MG 1-10.37.2

*Locked KVA per horsepower range includes the lower figure up to, but not including, the higher figure. For example, 3.14 is designated by letter A and 3.15 by letter B.

$$\frac{\text{Start. Kva}}{\text{per Hp}} = \frac{\text{Volts} \times \text{Locked-rotor Amp}}{1000 \times \text{Horsepower}} \times \begin{cases} 1 \text{ for 1-phase} \\ 2 \text{ for 2-phase} \\ 1.732 \text{ for 3-ph} \end{cases}$$

Code Letters Usually Applied to Ratings of Motors Normally Started On Full Voltage

Code Letters		F	G	H	J	K	L
Horsepower	3-phase	15 up	10–7½	5	3	2–1½	1
	1-phase	—	5	3	2–1½	1–¾	½

Chart 30   NEMA code letters for A-C motors.

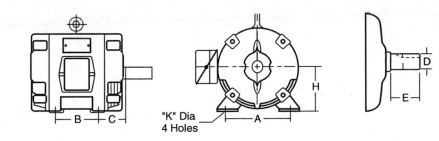

"K" Dia
4 Holes

*Letters in parentheses are NEMA equivalent to IEC.

Frame	H	(D)	A	(2E)	B	(2F)	C	(BA)	K	(H)	D	(U)	E	(N–W)
	mm	IN	mm	IN	mm	IN	mm	IN	mm	IN	mm	IN	mm	IN
56	56	2.2	90	3.54	71	2.8	36	1.42	5.8	.228	9	.354	20	.787
63	63	2.48	100	3.94	80	3.15	40	1.57	7	.276	11	.433	23	.905
71	71	2.8	112	4.41	90	3.54	45	1.77	7	.276	14	.551	30	1.18
80	80	3.15	125	4.92	100	3.94	50	19.7	10	.394	19	.748	40	1.57
90S	90	3.54	140	5.51	100	3.94	56	2.2	10	.394	24	.945	50	1.97
90L	90	3.54	140	5.51	125	4.92	56	2.2	10	.394	24	.945	50	1.97
100L	100	3.94	160	6.3	140	5.5	63	2.48	12	.472	28	1.1	60	2.36
112M	112	4.41	190	7.48	140	5.5	70	2.75	12	.472	28	1.1	60	2.36
132S	132	5.2	216	8.5	140	5.5	89	3.5	12	.472	38	1.5	80	3.15
132M	132	5.2	216	8.5	178	7	89	3.5	12	.472	38	1.5	80	3.15
160M	160	6.3	254	10	210	8.27	108	4.25	15	.591	Manufacturers do not agree			
160L	160	6.3	254	10	254	10	108	4.25	15	.591	beyond the 132 frame.			

Frame	H	(D)	A	(2E)	B	(2F)	C	(BA)	K	(H)
	mm	IN	mm	IN	mm	IN	mm	IN	mm	IN
180M	180	7.09	279	11	241	9.5	121	4.76	15	.591
180L	180	7.09	279	11	279	11	121	4.76	15	.591
200M	200	7.87	318	12.5	267	10.5	133	5.24	19	.748
200L	200	7.87	318	12.5	305	12	133	5.24	19	.748
225S	225	8.86	356	14	286	11.25	149	5.87	19	.748
225M	225	8.86	356	14	311	12.25	149	5.87	19	.748
250S	250	9.84	406	16	311	12.25	168	6.62	24	.945
250M	250	9.84	406	16	349	13.75	168	6.62	24	.945
280S	280	11	457	18	368	14.5	190	7.48	24	.945
280M	280	11	457	18	419	16.38	190	7.48	24	.945
315S	315	12.4	508	20	406	16	216	8.5	28	1.102
315M	315	12.4	508	20	457	18	216	8.5	28	1.102
315L	315	12.4	508	20	508	20	216	8.5	28	1.102
355S	355	14	610	24	500	19.69	254	10	28	1.102
355M	355	14	610	24	560	22	254	10	28	1.102
355L	355	14	610	24	630	24.8	254	10	28	1.102
400M	400	15.75	686	27	630	24.8	280	11	35	1.378
400L	400	15.75	686	27	710	27.95	280	11	35	1.378

IEC = International Electrotechnical Commission

Chart 31 Frame dimensions IEC—foot-mounted three-phase motors.

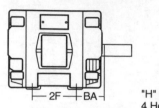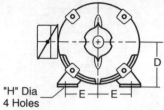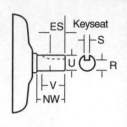

Frame	D	E	2F	H	U	BA	N-W	V min.	R	ES min.	S
48	3	2.12	2.75	.34	.5	2.5	1.5		.453		Flat
56	3.5	2.44	3	.34	.625	2.75	1.88		.517	1.41	.188
143	3.5	2.75	4	.34	.75	2.25	2	1.75	.643	1.41	.188
143T	3.5	2.75	4	.34	.875	2.25	2.25	2	.771	1.41	.188
145	3.5	2.75	5	.34	.75	2.25	2	1.75	.643	1.41	.188
145T	3.5	2.75	5	.34	.875	2.25	2.25	2	.771	1.41	.188
182	4.5	3.75	4.5	.41	.875	2.75	2.25	2	.771	1.41	.188
182T	4.5	3.75	4.5	.41	1.125	2.75	2.75	2.5	.986	1.78	.25
184	4.5	3.75	5.5	.41	.875	2.75	2.25	2	.771	1.41	.188
184T	4.5	3.75	5.5	.41	1.125	2.75	2.75	2.5	.986	1.78	.25
203	5	4	5.5	.41	.75	3.12	2.25	2	.643	1.53	.188
204	5	4	6.5	.41	.75	3.12	2.25	2	.643	1.53	.188
213	5.25	4.25	5.5	.41	1.125	3.5	3	2.75	.986	2.03	.25
213T	5.25	4.25	5.5	.41	1.375	3.5	3.38	3.12	1.201	2.41	.312
215	5.25	4.25	7	.41	1.125	3.5	3	2.75	.986	2.03	.25
215T	5.25	4.25	7	.41	1.375	3.5	3.38	3.12	1.201	2.41	.312
224	5.5	4.5	6.75	.41	1	3.5	3	2.75	.857	2.03	.25
225	5.5	4.5	7.5	.41	1	3.5	3	2.75	.857	2.03	.25
254	6.25	5	8.25	.53	1.125	4.25	3.37	3.12	.986	2.03	.25
254U	6.25	5	8.25	.53	1.375	4.25	3.75	3.5	1.201	2.78	.312
254T	6.25	5	8.25	.53	1.625	4.25	4	3.75	1.416	2.91	.375
256U	6.25	5	10	.53	1.375	4.25	3.75	3.5	1.201	2.78	.312
256T	6.25	5	10	.53	1.625	4.25	4	3.75	1.416	2.91	.375
284	7	5.5	9.5	.53	1.25	4.75	3.75	3.5	.986	2.03	.25
284U	7	5.5	9.5	.53	1.625	4.75	4.88	4.62	1.416	3.78	.375
284T	7	5.5	9.5	.53	1.875	4.75	4.62	4.38	1.591	3.28	.5
284TS	7	5.5	9.5	.53	1.625	4.75	3.25	3	1.416	1.91	.375
286U	7	5.5	11	.53	1.625	4.75	4.88	4.62	1.416	3.78	.375
286T	7	5.5	11	.53	1.875	4.75	4.62	4.38	1.591	3.28	.5
286TS	7	5.5	11	.53	1.625	4.75	3.25	3	1.416	1.91	.375
324	8	6.25	10.5	.66	1.625	5.25	4.87	4.62	1.416	3.78	.375
324U	8	6.25	10.5	.66	1.875	5.25	5.62	5.38	1.591	4.28	.5
324S	8	6.25	10.5	.66	1.625	5.25	3.25	3	1.416	1.91	.375
324T	8	6.25	10.5	.66	2.125	5.25	5.25	5	1.845	3.91	.5
324TS	8	6.25	10.5	.66	1.875	5.25	3.75	3.5	1.591	2.03	.5
326	8	6.25	12	.66	1.625	5.25	4.87	4.62	1.416	3.78	.375
326U	8	6.25	12	.66	1.875	5.25	5.62	5.38	1.591	4.28	.5
326S	8	6.25	12	.66	1.625	5.25	3.25	3	1.416	1.91	.375
326T	8	6.25	12	.66	2.125	5.25	5.25	5	1.845	3.91	.5
326TS	8	6.25	12	.66	1.875	5.25	3.75	3.5	1.591	2.03	.5

All dimensions in inches. Reference NEMA Standards MG 1-11.31.

Chart 32   NEMA frame dimensions—foot-mounted A-C motors and generators.

NEMA Shaft Dimensions
D-C Motors and Generators

Frame Range	Drive End—Belt						Drive End—Direct						Opp Drive End—Straight					
	U	N-W	V Min	R	ES Min	S	U	N-W	V Min	R	ES Min	S	FU	FN-FW	FV Min	FR	FES Min	FS
42	.375	1.12		.328		Flat	.375	1.12		.328		Flat						
48	.5	1.5		.453		Flat	.5	1.5		.453		Flat						
56	.625	1.88		.517	1.41	.188	.625	1.88		.517	1.41	.188						
56H	.625	1.88		.517	1.41	.188	.625	1.88		.517	1.41	.188						
142AT-1412AT	.875	1.75	1.5	.771	.91	.188							.625	1.25	1	.517	.66	.188
162AT-1610AT	.875	1.75	1.5	.771	.91	.188							.625	1.25	1	.517	.66	.188
182AT-1810AT	1.125	2.25	2	.986	1.41	.25							.875	1.75	1.5	.771	.91	.188
213AT-2110AT	1.375	2.75	2.5	1.201	1.78	.312							1.125	2.25	2	.986	1.41	.25
253AT-259AT	1.625	3.25	3	1.416	2.28	.375							1.375	2.75	2.5	1.201	1.78	.312
283AT-289AT	1.875	3.75	3.5	1.591	2.53	.5							1.625	3.25	3	1.416	2.28	.375
323AT-329AT	2.125	4.25	4	1.845	3.03	.5							1.875	3.75	3.5	1.591	2.53	.5
363AT-369AT	2.375	4.75	4.5	2.021	3.53	.625							2.125	4.25	4	1.845	3.03	.5
403AT-409AT	2.625	5.25	5	2.275	4.03	.625							2.375	4.75	4.5	2.021	3.53	.625
443AT-449AT	2.875	5.75	5.5	2.45	4.53	.75							2.625	5.25	5	2.275	4.03	.625
502AT-509AT	3.25	6.5	6.25	2.831	5.28	.75							2.875	5.75	5.5	2.45	4.53	.75
583A-588A	3.25	9.75	9.5	2.831	8.28	.75	2.875	5.75	5.5	2.45	4.28	.75						
683A-688A	3.625	10.88	10.62	3.134	9.53	.875	3.25	6.5	6.25	2.831	5.03	.75						

All dimensions in inches. Reference NEMA Standards MG 1-11.60, 11.61 & 11.62.

Chart 33   NEMA shaft dimensions—D-C motors and generators.

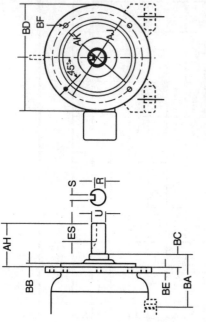

NEMA Frame Dimensions
Type D Flange-Mounting Foot or Footless A-C Motors

Frame Designations	AJ	AK	BA	BB	BC	BD Max	BE Nom	BF Hole Number	Size	Recommended Bolt Length	U	AH	R	Keyseat ES Min	S
143TD and 145TD	10.00	9.000	2.75	0.25	0.00	11.00	0.50	4	0.53	1.25	0.8750	2.25	0.771	1.41	0.188
182D and 184TD	10.00	9.000	3.50	0.25	0.00	11.00	0.50	4	0.53	1.25	1.1250	2.75	0.986	1.78	0.250
182TD and 215TD	10.00	9.000	4.25	0.25	0.00	11.00	0.50	4	0.53	1.25	1.3750	3.38	1.201	2.41	0.312
213D and 256TD	12.50	11.000	4.75	0.25	0.00	14.00	0.75	4	0.81	2.00	1.625	4.00	1.416	2.91	0.375
234TD and 286TD	12.50	11.000	4.75	0.25	0.00	14.00	0.75	4	0.81	2.00	1.875	4.62	1.591	3.28	0.500
284TSD and 286TSD	12.50	11.000	4.75	0.25	0.00	14.00	0.75	4	0.81	2.00	1.625	3.25	1.416	1.91	0.375
324SD and 326TD	16.00	14.000	5.25	0.25	0.00	18.00	0.75	4	0.81	2.00	2.125	5.25	1.845	3.91	0.500
284TSD and 326TSD	16.00	14.000	5.25	0.25	0.00	18.00	0.75	4	0.81	2.00	1.875	3.75	1.591	2.03	0.500
364TD and 365TD	16.00	14.000	5.88	0.25	0.00	18.00	0.75	4	0.81	2.00	2.375	5.88	2.021	4.28	0.625
364TSD and 365TSD	16.00	14.000	5.88	0.25	0.00	18.00	0.75	4	0.81	2.00	1.875	3.75	1.591	2.03	0.500
404TD and 405TD	20.00	18.000	6.62	0.25	0.00	22.00	1.00	8	0.81	2.25	2.875	7.25	2.450	5.65	0.750
404TSD and 405TSD	20.00	18.000	6.62	0.25	0.00	22.00	1.00	8	0.81	2.25	2.125	4.25	1.845	2.78	0.500
444TD and 445TD	20.00	18.000	7.50	0.25	0.00	22.00	1.00	8	0.81	2.25	3.375	8.50	2.880	6.91	0.875
444TSD and 445TSD	20.00	18.000	7.50	0.25	0.00	22.00	1.00	8	0.81	2.25	2.375	4.75	2.021	3.03	0.625
504SD and 505SD frame series	22.00	18.000	—	0.25	0.00	25.00	—	8	0.81	—	2.125	4.25	1.82	—	.500

Dimensions in inches.
Reference NEMA Standards MG 1-11.36.

Chart 34 NEMA frame dimensions.

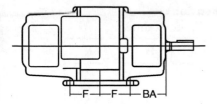

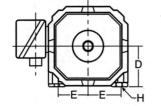

NEMA Frame Dimensions
D-C Motors and Generators, Foot-Mounted

Frame	D	E	2F	H	BA
42	2.62	1.75	1.69	.28	2.06
48	3	2.12	2.75	.34	2.5
56	3.5	2.44	3	.34	2.75
56H	3.5	2.44	5	.34	2.75
142AT	3.5	2.75	3.5	.34	2.75
143AT	3.5	2.75	4	.34	2.75
144AT	3.5	2.75	4.5	.34	2.75
145AT	3.5	2.75	5	.34	2.75
146AT	3.5	2.75	5.5	.34	2.75
147AT	3.5	2.75	6.25	.34	2.75
148AT	3.5	2.75	7	.34	2.75
149AT	3.5	2.75	8	.34	2.75
1410AT	3.5	2.75	9	.34	2.75
1411AT	3.5	2.75	10	.34	2.75
1412AT	3.5	2.75	11	.34	2.75
162AT	4	3.12	4	.41	2.5
163AT	4	3.12	4.5	.41	2.5
164AT	4	3.12	5	.41	2.5
165AT	4	3.12	5.5	.41	2.5
166AT	4	3.12	6.25	.41	2.5
167AT	4	3.12	7	.41	2.5
168AT	4	3.12	8	.41	2.5
169AT	4	3.12	9	.41	2.5
1610AT	4	3.12	10	.41	2.5
182AT	4.5	3.75	4.5	.41	2.75
183AT	4.5	3.75	5	.41	2.75
184AT	4.5	3.75	5.5	.41	2.75
185AT	4.5	3.75	6.25	.41	2.75
186AT	4.5	3.75	7	.41	2.75
187AT	4.5	3.75	8	.41	2.75
188AT	4.5	3.75	9	.41	2.75
189AT	4.5	3.75	10	.41	2.75
1810AT	4.5	3.75	11	.41	2.75
213AT	5.25	4.25	5.5	.41	3.5
214AT	5.25	4.25	6.25	.41	3.5
215AT	5.25	4.25	7	.41	3.5
216AT	5.25	4.25	8	.41	3.5
217AT	5.25	4.25	9	.41	3.5
218AT	5.25	4.25	10	.41	3.5
219AT	5.25	4.25	11	.41	3.5
2110AT	5.25	4.25	12.5	.41	3.5
253AT	6.25	5	7	.53	4.25
254AT	6.25	5	8.25	.53	4.25
255AT	6.25	5	9	.53	4.25
256AT	6.25	5	10	.53	4.25
257AT	6.25	5	11	.53	4.25

All dimensions in inches. Reference NEMA Standards MG 1-11.60, 11.61 & 11.62.

Chart 35 NEMA frame dimensions—D-C motors and generators, foot mounted.

NEMA Frame Dimensions
D-C Motors and Generators, Foot-Mounted

Frame	D	E	2F	H	BA
258AT	6.25	5	12.5	.53	4.25
259AT	6.25	5	14	.53	4.25
283AT	7	5.5	8	.53	4.75
284AT	7	5.5	9.5	.53	4.75
285AT	7	5.5	10	.53	4.75
286AT	7	5.5	11	.53	4.75
287AT	7	5.5	12.5	.53	4.75
288AT	7	5.5	14	.53	4.75
289AT	7	5.5	16	.53	4.75
323AT	8	6.25	9	.66	5.25
324AT	8	6.25	10.5	.66	5.25
325AT	8	6.25	11	.66	5.25
326AT	8	6.25	12	.66	5.25
327AT	8	6.25	14	.66	5.25
328AT	8	6.25	16	.66	5.25
329AT	8	6.25	18	.66	5.25
363AT	9	7	10	.81	5.88
364AT	9	7	11.25	.81	5.88
365AT	9	7	12.25	.81	5.88
366AT	9	7	14	.81	5.88
367AT	9	7	16	.81	5.88
368AT	9	7	18	.81	5.88
369AT	9	7	20	.81	5.88
403AT	10	8	11	.94	6.62
404AT	10	8	12.25	.94	6.62
405AT	10	8	13.75	.94	6.62
406AT	10	8	16	.94	6.62
407AT	10	8	18	.94	6.62
408AT	10	8	20	.94	6.62
409AT	10	8	22	.94	6.62
443AT	11	9	12.5	1.06	7.5
444AT	11	9	15	1.06	7.5
445AT	11	9	16.5	1.06	7.5
446AT	11	9	18	1.06	7.5
447AT	11	9	20	1.06	7.5
448AT	11	9	22	1.06	7.5
449AT	11	9	25	1.06	7.5
502AT	12.5	10	12.5	1.19	8.5
503AT	12.5	10	14	1.19	8.5
504AT	12.5	10	16	1.19	8.5
505AT	12.5	10	18	1.19	8.5
506AT	12.5	10	20	1.19	8.5
507AT	12.5	10	22	1.19	8.5
508AT	12.5	10	25	1.19	8.5

All dimensions in inches. Reference NEMA Standards MG 1-11.60, 11.61 & 11.62.

Chart 36 NEMA frame dimensions—D-C motors and generators, foot mounted.

NEMA Frame Dimensions
D-C Motors and Generators, Foot-Mounted

Frame	D	E	2F	H	BA
509AT	12.5	10	28	1.19	8.5
583A	14.5	11.5	16	1.19	10
584A	14.5	11.5	18	1.19	10
585A	14.5	11.5	20	1.19	10
586A	14.5	11.5	22	1.19	10
587A	14.5	11.5	25	1.19	10
588A	14.5	11.5	28	1.19	10
683A	17	13.5	20	1.19	11.5
684A	17	13.5	22	1.19	11.5
685A	17	13.5	25	1.19	11.5
686A	17	13.5	28	1.19	11.5
687A	17	13.5	32	1.19	11.5
688A	17	13.5	36	1.19	11.5

All dimensions in inches. Reference NEMA Standards MG 1-11.60, 11.61 & 11.62.

Chart 36   *Continued.*

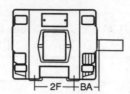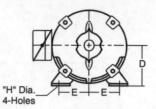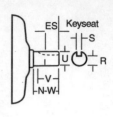

Frame	D	E	2F	H	U	BA	N-W	V min.	R	ES min.	S
364	9	7	11.25	.66	1.875	5.88	5.62	5.38	1.591	4.28	.5
364A	9	7	11.25	.66	1.625	5.88	3.25	3	1.416	1.91	.375
364U	9	7	11.25	.66	2.125	5.88	6.37	6.12	1.845	5.03	.5
364US	9	7	11.25	.66	1.875	5.88	3.75	3.5	1.591	2.03	.5
364T	9	7	11.25	.66	2.375	5.88	5.88	5.62	2.01	4.28	.625
364TS	9	7	11.25	.66	1.875	5.88	3.75	3.5	1.591	2.03	.5
365	9	7	12.25	.66	1.875	5.88	5.62	5.38	1.591	4.28	.5
365S	9	7	12.25	.66	1.625	5.88	3.25	3	1.416	1.91	.375
365U	9	7	12.25	.66	2.125	5.88	6.37	6.12	1.845	5.03	.5
365US	9	7	12.25	.66	1.875	5.88	3.75	3.5	1.591	2.03	.5
365T	9	7	12.25	.66	2.375	5.88	5.88	5.26	2.021	4.28	.625
365TS	9	7	12.25	.66	1.875	5.88	3.75	3.5	1.591	2.03	.5
404	10	8	12.25	.81	2.125	6.62	6.37	6.12	1.845	5.03	.5
404S	10	8	12.25	.81	1.875	6.62	3.75	3.5	1.591	2.03	.5
404U	10	8	12.25	.81	2.375	6.62	`7.12	6.88	2.021	5.53	.625
404US	10	8	12.25	.81	2.125	6.62	4.25	4	1.845	2.78	.5
404T	10	8	12.25	.81	2.875	6.62	7.25	7	2.45	5.65	.75
404TS	10	8	12.25	.81	2.125	6.62	4.25	4	1.845	2.78	.5
405	10	8	13.75	.81	2.125	6.62	6.37	6.12	1.845	5.03	.5
405S	10	8	13.75	.81	1.875	6.62	3.75	3.5	1.591	2.03	.5
405U	10	8	13.75	.81	2.375	6.62	7.12	6.88	2.021	5.53	.625
405US	10	8	13.75	.81	2.125	6.62	4.25	4	1.845	2.78	.5
405T	10	8	13.75	.81	2.875	6.62	7.25	7	2.45	5.65	.75
405TS	10	8	13.75	.81	2.125	6.62	4.25	4	1.845	2.78	.5
444	11	9	14.5	.81	2.375	7.5	7.12	6.88	2.021	5.53	.625
444S	11	9	14.5	.81	2.125	7.5	4.25	4	1.845	2.78	.5
444U	11	9	14.5	.81	2.875	7.5	8.62	8.38	2.45	7.03	.75
444US	11	9	14.5	.81	2.125	7.5	4.25	4	1.845	2.78	.5
444T	11	9	14.5	.81	3.375	7.5	8.5	8.25	2.88	6.91	.875
444TS	11	9	14.5	.81	2.375	7.5	4.75	4.5	2.021	3.03	.625
445	11	9	16.5	.81	2.375	7.5	7.12	6.88	2.021	5.53	.625
445S	11	9	16.5	.81	2.125	7.5	4.25	4	1.845	2.78	.5
445U	11	9	16.5	.81	2.875	7.5	8.62	8.38	2.45	7.03	.75
445US	11	9	16.5	.81	2.125	7.5	4.25	4	1.845	2.78	.5
445G	11	9	16.5	.81	3.375	7.5	8.5	8.25	2.88	6.91	.875
445TS	11	9	16.5	.81	2.375	7.5	4.75	4.5	2.021	3.03	.625
447TS	11	9	20	Dimensions vary with manufacturers							
449TS	11	9	25	Dimensions vary with manufacturers							
504U	12.5	10	16	.94	2.875	8.5	8.62	8.38	2.45	7.28	.75
505	12.5	10	18	.94	2.875	8.5	8.62	8.38	2.45	7.28	.75
505S	12.5	10	18	.94	2.125	8.5	4.25	4	1.845	2.78	.5

All dimensions in inches. Reference NEMA Standards MG 1-11.31.

Chart 37 NEMA frame dimensions—foot-mounted A-C motors and generators.

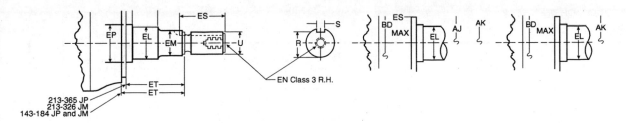

213-365 JP
213-326 JM
143-184 JP and JM

EN Class 3 R.H.

Type JM and JP Face Mounting
Closed–Coupled, AC Pump Motors

Frame Designations	U	AH	AJ	AK	BB	BD Max	BF		
							Number	Tap Size	Bolt Penetration Allowance
143JM and 145JM	0.8745 / 0.8740	4.281 / 4.219	5.875	4.500 / 4.497	0.156 / 0.125	6.62	4	$3/8$–16	0.56
143JP and 145JP	0.8745 / 0.8740	7.343 / 7.281	5.875	4.500 / 4.497	0.156 / 0.125	6.62	4	$3/8$–16	0.56
182JM and 184JM	0.8745 / 0.8740	4.281 / 4.219	5.875	4.500 / 4.497	0.156 / 0.125	6.62	4	$3/8$–16	0.56
182JP and 184JP	0.8745 / 0.8740	7.343 / 7.281	5.875	4.500 / 4.497	0.156 / 0.125	6.62	4	$3/8$–16	0.56
213JM and 215JM	0.8745 / 0.8740	4.281 / 4.219	7.250	8.500 / 8.497	0.312 / 0.250	9.00	4	$1/2$–13	0.75
213JP and 215JP	1.2495 / 1.2490	8.156 / 8.094	7.250	8.500 / 8.497	0.312 / 0.250	9.00	4	$1/2$–13	0.75
254JM and 256JM	1.2495 / 1.2490	5.281 / 5.219	7.250	8.500 / 8.497	0.312 / 0.250	10.00	4	$1/2$–13	0.75
254JP and 256JP	1.2495 / 1.2490	8.156 / 8.094	7.250	8.500 / 8.497	0.312 / 0.250	10.00	4	$1/2$–13	0.75
284JM and 286JM	1.2495 / 1.2490	5.281 / 5.219	11.000	12.500 / 12.495	0.312 / 0.250	14.00	4	$5/8$–11	0.94
284JP and 286JP	1.2495 / 1.2490	8.156 / 8.094	11.000	12.500 / 12.495	0.312 / 0.250	14.00	4	$5/8$–11	0.94
324JM and 326JM	1.2495 / 1.2490	5.281 / 5.219	11.000	12.500 / 12.495	0.312 / 0.250	14.00	4	$5/8$–11	0.94
324JP and 326JP	1.2495 / 1.2490	8.156 / 8.094	11.000	12.500 / 12.495	0.312 / 0.250	14.00	4	$5/8$–11	0.94
364JP and 365JP	1.6245 / 1.6240	8.156 / 8.094	11.000	12.500 / 12.495	0.312 / 0.250	14.00	4	$5/8$–11	0.94

All dimensions in inches. Reference NEMA Standards MG 1-18.614.

Chart 38 NEMA frame dimensions.

Type JM and JP Face Mounting
Closed–Coupled, AC Pump Motors

Frame Designations	EL	EM	EN Tap Size	EN Tap Drill Depth Max	EN Bolt Penetration Allowance	EP Min	EQ	ER Min	R	ES Min	S	ET
143JM and 145JM	1.156 / 1.154	1.0000 / 0.9995	3/8–16	1.12	0.75	1.156	0.640 / 0.610	4.25	0.771–0.756	1.65	0.190–0.188	2.890 / 2.860
143JP and 145JP	1.156 / 1.154	1.0000 / 0.9995	3/8–16	1.12	0.75	1.156	1.578 / 1.548	7.312	0.771–0.756	1.65	0.190–0.188	5.952 / 5.922
182JM and 184JM	1.250 / 1.248	1.0000 / 0.9995	3/8–16	1.12	0.75	1.250	0.640 / 0.610	4.25	0.771–0.756	1.65	0.190–0.188	2.890 / 2.860
182JP and 184JP	1.250 / 1.248	1.0000 / 0.9995	3/8–16	1.12	0.75	1.250	1.578 / 1.548	7.312	0.771–0.756	1.65	0.190–0.188	5.952 / 5.922
213JM and 215JM	1.250 / 1.248	1.0000 / 0.9995	3/8–16	1.12	0.75	1.750	0.640 / 0.610	4.25	0.771–0.756	1.65	0.190–0.188	2.890 / 2.860
213JP and 215JP	1.750 / 1.748	1.3750 / 1.3745	1/2–13	1.50	1.00	1.750	2.390 / 2.370	8.125	1.112–1.097	2.53	0.252–0.250	5.890 / 5.860
254JM and 256JM	1.750 / 1.748	1.3750 / 1.3745	1/2–13	1.50	1.00	1.750	0.640 / 0.610	5.25	1.112–1.097	2.53	0.252–0.250	3.015 / 2.985
254JP and 256JP	1.750 / 1.748	1.3750 / 1.3745	1/2–13	1.50	1.00	1.750	2.390 / 2.360	8.125	1.112–1.097	2.53	0.252–0.250	5.890 / 5.860
284JM and 286JM	1.750 / 1.748	1.3750 / 1.3745	1/2–13	1.50	1.00	2.125	0.645 / 0.605	5.25	1.112–1.097	2.53	0.252–0.250	3.020 / 2.980
284JP and 286JP	1.750 / 1.748	1.3750 / 1.3745	1/2–13	1.50	1.00	2.125	2.390 / 2.360	8.125	1.112–1.097	2.53	0.252–0.250	5.895 / 5.855
324JM and 326JM	1.750 / 1.748	1.3750 / 1.3745	1/2–13	1.50	1.00	2.125	0.645 / 0.605	5.25	1.112–1.097	2.53	0.252–0.250	3.020 / 2.980
324JP and 326JP	1.750 / 1.748	1.3750 / 1.3745	1/2–13	1.50	1.00	2.125	2.395 / 2.355	8.125	1.112–1.097	2.53	0.252–0.250	5.895 / 5.855
364JP and 365JP	2.125 / 2.123	1.7500 / 1.7495	1/2–13	1.50	1.00	2.500	2.395 / 2.355	8.125	1.416–1.401	2.53	0.377–0.375	5.895 / 5.855

All dimensions in inches. Reference NEMA Standards MG 1–18.614.

Chart 39 NEMA frame dimensions.

NEMA Frame Dimensions
Type C Face–Mounting Foot or Footless A-C Motors

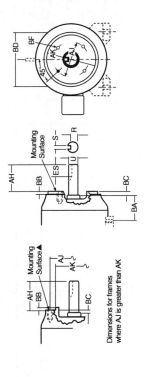

Dimensions for frames where AJ is greater than AK

Frame Designations	AJ	AK	BA	BB Min	BC	BD Max	BF Hole Number	BF Hole Tap Size	BF Hole Bolt Penetration Allowance	U	AH	Keyseat R	Keyseat ES Min	S
42C	3.750	3.000	2.062	0.16	−0.19	5.00	4	1/4–20	—	0.3750	1.312	0.328	—	flat
48C	3.750	3.000	2.50	0.16	−0.19	5.625	4	1/4–20	—	0.500	1.69	0.453	—	flat
56C	5.875	4.500	2.75	0.16	−0.19	6.50	4	3/8–16	—	0.6250	2.06	0.517	1.41	0.188
143TC and 145TC	5.875	4.500	2.75	0.16	+0.12	6.50	4	3/8–16	0.56	0.8750	2.12	0.771	1.41	0.188
182C and 184C	7.250	8.500	3.50	0.25	+0.12	9.00	4	1/2–13	0.75	1.1250	2.62	0.986	1.78	0.250
182TCH and 184TCH	5.875	4.500	3.50	0.16	+0.12	6.50	4	3/8–16	0.56	1.1250	2.62	0.986	1.78	0.250
213C and 215C	7.250	8.500	4.25	0.25	+0.25	9.00	4	1/2–13	0.75	1.3750	3.12	1.201	2.41	0.312
254UC and 256UC	7.250	8.500	4.75	0.25	+0.25	10.00	4	1/2–13	0.75	1.625	3.75	1.416	2.91	0.375
284TC and 286TC	9.000	10.500	4.75	0.25	+0.25	11.25	4	1/2–13	0.75	1.875	4.38	1.519	3.28	0.500
284TSC and 286TSC	9.000	10.500	4.75	0.25	+0.25	11.25	4	1/2–13	0.75	1.625	3.00	1.416	1.91	0.375
324UC and 326UC	11.000	12.500	5.25	0.25	+0.25	14.00	4	5/8–11	0.94	2.125	5.00	1.845	3.91	0.500
324TSC and 326TSC	11.000	12.500	5.25	0.25	+0.25	14.00	4	5/8–11	0.94	1.875	3.50	1.591	2.03	0.500
364TC and 365TC	11.000	12.500	5.88	0.25	+0.25	14.00	8	5/8–11	0.94	2.375	5.62	2.021	4.28	0.625
364TSC and 365TSC	11.000	12.500	5.88	0.25	+0.25	14.00	8	5/8–11	0.94	1.875	3.50	1.591	2.03	0.500
404TC and 405TC	11.000	12.500	6.62	0.25	+0.25	15.50	8	5/8–11	0.94	2.875	7.00	2.450	5.65	0.750
404TSC and 405TSC	11.000	12.500	6.62	0.25	+0.25	15.50	8	5/8–11	0.94	2.125	4.00	1.845	2.78	0.500
444TC and 445TC	14.000	16.000	7.50	0.25	+0.25	18.00	8	5/8–11	0.94	3.375	8.25	2.880	6.91	0.875
444TSC and 445TSC	14.000	16.000	7.50	0.25	+0.25	18.00	8	5/8–11	0.94	2.375	4.50	2.021	3.03	0.625
504SC and 505SC frame series	14.500	16.500	—	0.25	+0.25	18.00	4	5/8–11	0.94	2.125	4.00	1.845	—	.500

Dimensions in inches. Reference NEMA Standards MG 1–11.34.

Chart 40 NEMA frame dimensions—type C face-mounting foot or slotless A-C motors.

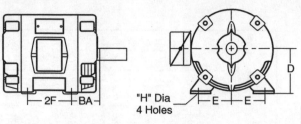

| | | | | | | | | | | Keyseat | | |
Frame	D	E	2F	H	U	BA	N-W	V min.	R	ES min.	S
48	3	2.12	2.75	.34	.5	2.5	1.5		.453		Flat
56	3.5	2.44	3	.34	.625	2.75	1.88		.517	1.41	.188
143	3.5	2.75	4	.34	.75	2.25	2	1.75	.643	1.41	.188
143T	3.5	2.75	4	.34	.875	2.25	2.25	2	.771	1.41	.188
145	3.5	2.75	5	.34	.75	2.25	2	1.75	.643	1.41	.188
145T	3.5	2.75	5	.34	.875	2.25	2.25	2	.771	1.41	.188
182	4.5	3.75	4.5	.41	.875	2.75	2.25	2	.771	1.41	.188
182T	4.5	3.75	4.5	.41	1.125	2.75	2.75	2.5	.986	1.78	.25
184	4.5	3.75	5.5	.41	.875	2.75	2.25	2	.771	1.41	.188
184T	4.5	3.75	5.5	.41	1.125	2.75	2.75	2.5	.986	1.78	.25
203	5	4	5.5	.41	.75	3.12	2.25	2	.643	1.53	.188
204	5	4	6.5	.41	.75	3.12	2.25	2	.643	1.53	.188
213	5.25	4.25	5.5	.41	1.125	3.5	3	2.75	.986	2.03	.25
213T	5.25	4.25	5.5	.41	1.375	3.5	3.38	3.12	1.201	2.41	.312
215	5.25	4.25	7	.41	1.125	3.5	3	2.75	.986	2.03	.25
215T	5.25	4.25	7	.41	1.375	3.5	3.38	3.12	1.201	2.41	.312
224	5.5	4.5	6.75	.41	1	3.5	3	2.75	.857	2.03	.25
225	5.5	4.5	7.5	.41	1	3.5	3	2.75	.857	2.03	.25
254	6.25	5	8.25	.53	1.125	4.25	3.37	3.12	.986	2.03	.25
254U	6.25	5	8.25	.53	1.375	4.25	3.75	3.5	1.201	2.78	.312
254T	6.25	5	8.25	.53	1.625	4.25	4	3.75	1.416	2.91	.375
256U	6.25	5	10	.53	1.375	4.25	3.75	3.5	1.201	2.78	.312
256T	6.25	5	10	.53	1.625	4.25	4	3.75	1.416	2.91	.375
284	7	5.5	9.5	.53	1.25	4.75	3.75	3.5	.986	2.03	.25
284U	7	5.5	9.5	.53	1.625	4.75	4.88	4.62	1.416	3.78	.375
284T	7	5.5	9.5	.53	1.875	4.75	4.62	4.38	1.591	3.28	.5
284TS	7	5.5	9.5	.53	1.625	4.75	3.25	3	1.416	1.91	.375
286U	7	5.5	11	.53	1.625	4.75	4.88	4.62	1.416	3.78	.375
286T	7	5.5	11	.53	1.875	4.75	4.62	4.38	1.591	3.28	.5
286TS	7	5.5	11	.53	1.625	4.75	3.25	3	1.416	1.91	.375
324	8	6.25	10.5	.66	1.625	5.25	4.87	4.62	1.416	3.78	.375
324U	8	6.25	10.5	.66	1.875	5.25	5.62	5.38	1.591	4.28	.5
324S	8	6.25	10.5	.66	1.625	5.25	3.25	3	1.416	1.91	.375
324T	8	6.25	10.5	.66	2.125	5.25	5.25	5	1.845	3.91	.5
324TS	8	6.25	10.5	.66	1.875	5.25	3.75	3.5	1.591	2.03	.5
326	8	6.25	12	.66	1.625	5.25	4.87	4.62	1.416	3.78	.375
326U	8	6.25	12	.66	1.875	5.25	5.62	5.38	1.591	4.28	.5
326S	8	6.25	12	.66	1.625	5.25	3.25	3	1.416	1.91	.375
326T	8	6.25	12	.66	2.125	5.25	5.25	5	1.845	3.91	.5
326TS	8	6.25	12	.66	1.875	5.25	8.75	3.5	1.591	2.03	.5

All dimensions in inches. Reference NEMA Standards MG 1-11.31.

Chart 41   NEMA frame dimensions—foot-mounted AC motors and generators.

Three-Phase Frame Sizes
TEFC Motors—General Purpose

RPM	3600			1800			1200			900		
NEMA Program	Orig.	1952 Rerate	1964 Rerate	Orig.	1952 Rerate	1964 Rerate	Orig.	1952 Rerate	1964 Rerate	Orig.	1952 Rerate	1964 Rerate
HP												
1	—	—	—	203	182	143T	204	184	145T	225	213	182T
1½	203	182	143T	204	184	145T	224	184	182T	254	213	184T
2	204	184	145T	224	184	145T	225	213	184T	254	215	213T
3	224	184	182T	225	213	182T	254	215	213T	284	254U	215T
5	225	213	184T	254	215	184T	284	254U	215T	324	256U	254T
7½	254	215	213T	284	254U	213T	324	256U	254T	325	284U	256T
10	284	254U	215T	324	256U	215T	326	284U	256T	364	286U	284T
15	324	256U	254T	326	284U	254T	364	324U	284T	365	326U	286T
20	326	286U	256T	364	286U	256T	365	326U	286T	404	364U	324T
25	365S	324U	284TS	365	324U	284T	404	364U	324T	405	365U	326T
30	404S	326S	286TS	404	326U	286T	405	365U	326T	444	404U	364T
40	405S	364US	324TS	405	364U	324T	444	404U	364T	445	405U	365T
50	444S	365US	362TS	444S	365US	326T	445	405U	365T	504U	444U	404T
60	445S	405US	364TS	445S	405US	364TS	504U	444U	404T	505	445U	405T
75	504S	444US	365TS	504S	444US	365TS	505	445U	405T	—	—	444T
100	505S	445US	405TS	505S	445US	405TS	—	—	444T	—	—	445T
125	—	—	444TS	—	—	444TS	—	—	445T	—	—	—
150	—	—	445TS	—	—	445TS	—	—	—	—	—	—

Chart 42 Three-phase frame sizes—TEFC motors, general purpose.

Three-Phase Frame Sizes
Open Motors—General Purpose

RPM	3600			1800			1200			900		
NEMA Program HP	Orig.	1952 Rerate	1964 Rerate	Orig.	1952 Rerate	1964 Rerate	Orig.	1952 Rerate	1964 Rerate	Orig.	1952 Rerate	1964 Rerate
30	364S	324S	284TS	365	326U	286T	405	365U	326T	444	404U	364T
40	365S	326S	286TS	404	364U	324T	444	404U	364T	445	405U	365T
50	404S	364US	324TS	405S	365US	326T	445	405U	365T	504U	444U	404T
60	405S	365US	326TS	444S	404US	364TS	504U	444U	404T	505	445U	405T
75	444S	404US	364TS	445S	405US	365TS	505	445U	405T	—	—	444T
100	445S	405US	365TS	504S	444US	404TS	—	—	444T	—	—	445T
125	504S	444US	404TS	505S	445US	405TS	—	—	445T	—	—	—
150	505S	445US	405TS	—	—	444TS	—	—	—	—	—	—
200	—	—	444TS	—	—	445TS	—	—	—	—	—	—
250	—	—	445TS	—	—	—	—	—	—	—	—	—

Chart 42 *Continued.*

NEMA Frame Assignments

Three-Phase Open Motors—General Purpose

NEMA Program HP	3600 RPM			1800 RPM			1200 RPM			900 RPM		
	Orig. Rerate	1952 Rerate	1964 Rerate	Orig.	1952 Rerate	1964 Rerate	Orig.	1952 Rerate	1964	Orig. Rerate	1952 Rerate	1964
1	—	—	—	203	182	143T	204	184	145T	225	213	182T
1.5	203	182	143T	204	184	145T	224	184	182T	254	213	184T
2	204	184	145T	224	184	145T	225	213	184T	254	215	213T
3	224	184	145T	225	213	182T	254	215	213T	284	254U	215T
5	225	213	182T	254	215	184T	284	254U	215T	324	256U	254T
7.5	254	215	184T	284	254U	213T	324	256U	254T	326	284U	256T
10	284	254U	213T	324	256U	215T	326	284U	256T	364	286U	284T
15	324	256U	215T	326	284U	254T	364	324U	284T	365	326U	286T
20	326	284U	254T	364	286U	256T	365	326U	286T	404	364U	324T
25	364S	286U	256T	364	324U	284T	404	364U	324T	405	365U	326T
30	364S	324S	284TS	365	326U	286T	405	365U	326T	444	404U	364T
40	365S	326S	286TS	404	364U	324T	444	404U	364T	445	405U	365T
50	404S	364US	324TS	405S	365US	326T	445	405U	365T	504U	444U	404T
60	405S	365US	326TS	444S	404US	364TS†	504U	444U	404T	505	445U	405T
75	444S	404US	364TS	445S	405US	365TS†	505	445U	405T	—	—	444T
100	445S	405US	365TS	504S	444US	404TS†	—	—	444T	—	—	445T
125	504S	444US	404TS	505S	445US	405TS†	—	—	445T	—	—	—
150	505S	445US	405TS	—	—	444TS†	—	—	—	—	—	—
200	—	—	444TS	—	—	445TS†	—	—	—	—	—	—
250	—	—	445TS	—	—	—	—	—	—	—	—	—

†When motors are to be used with V-belts or chain drive, the correct frame size is the one shown but with the suffix letter S omitted. For the corresponding shaft extension dimensions, see pages 303 and 304, Foot Mounted AC Machines.

Chart 43 NEMA frame assignments—Three-phase open motors, general purpose.

NEMA Frame Assignments

Three-Phase TEFC Motors—General Purpose

NEMA Program HP	3600 RPM			1800 RPM			1200 RPM			900 RPM		
	Orig. Rerate	1952 Rerate	1964 Rerate	Orig.	1952 Rerate	1964 Rerate	Orig.	1952 Rerate	1964	Orig. Rerate	1952 Rerate	1964
1	—	—	—	203	182	143T	204	184	145T	225	213	182T
1.5	203	182	143T	204	184	145T	224	184	182T	254	213	184T
2	204	184	145T	224	184	145T	225	213	184T	254	215	213T
3	224	184	182T	225	213	182T	254	215	213T	284	254U	215T
5	225	213	184T	254	215	184T	284	254U	215T	324	256U	254T
7.5	254	215	213T	284	254U	213T	324	256U	254T	326	284U	256T
10	284	254U	215T	324	256U	215T	326	284U	256T	364	286U	284T
15	324	256U	254T	326	284U	254T	364	324U	284T	365	326U	286T
20	326	286U	256T	364	286U	256T	365	326U	286T	404	364U	324T
25	365S	324U	284TS	365	324U	284T	404	364U	324T	405	365U	326T
30	404S	326S	286TS	404	326U	286T	405	365U	326T	444	404U	364T
40	405S	364US	324TS	405	364U	324T	444	404U	364T	445	405U	365T
50	444S	365US	326TS	444S	365US	326T	445	405U	365T	504U	444U	404T
60	445S	405US	364TS	445S	405US	364TS†	504U	444U	404T	505	445U	405T
75	504S	444US	365TS	504S	444US	365TS†	505	445U	405T	—	—	444T
100	505S	445US	405TS	505S	445US	405TS†	—	—	444T	—	—	445T
125	—	—	444TS	—	—	444TS†	—	—	445T	—	—	—
150	—	—	445TS	—	—	445TS†	—	—	—	—	—	—

†When motors are to be used with V-belt or chain drives, the correct frame size is the frame size shown but with the suffix letter S omitted. For the corresponding shaft extension dimensions, see pages 303 and 304, Foot Mounted AC Machines.

Chart 44   NEMA frame assignments—Three-phase TEFC motors, general purpose.

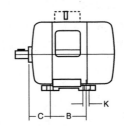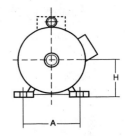

IEC Mounting Dimensions

Foot-Mounted AC Machines Dimensions for Mounting*

*Dimensions in inches

Frame Number	H	A	B	C	K	Bolt or Screw
56 M	2.20	3.55	2.80	1.40	0.23	M5
63 M	2.48	3.95	3.15	1.55	0.28	M6
71 M	2.79	4.40	3.55	1.75	0.28	M6
80 M	3.14	4.90	3.95	1.95	0.40	M8
90 S	3.54	5.50	3.95	2.20	0.40	M8
90 L	3.54	5.50	4.90	2.20	0.40	M8
100 S	3.93	6.30	4.40	2.50	0.48	M10
100 L	3.93	6.30	5.50	2.50	0.48	M10
112 S	4.40	7.50	4.50	2.75	0.48	M10
112 M	4.40	7.50	5.50	2.75	0.48	M10
132 S	5.19	8.50	5.50	3.50	0.48	M10
132 M	5.19	8.50	7.00	3.50	0.48	M10
160 S	6.29	10.00	7.00	4.25	0.58	M12
160 M	6.29	10.00	8.25	4.25	0.58	M12
160 L	6.29	10.00	10.00	4.25	0.58	M12
180 S	7.08	11.00	8.00	4.75	0.58	M12
180 M	7.08	11.00	9.50	4.75	0.58	M12
180 L	7.08	11.00	11.00	4.75	0.58	M12
200 S	7.87	12.50	9.00	5.25	0.73	M16
200 M	7.87	12.50	10.50	5.25	0.73	M16
200 L	7.87	12.50	12.00	5.25	0.73	M16
225 S	8.85	14.00	11.25	5.85	0.73	M16
225 M	8.85	14.00	12.25	5.85	0.73	M16
250 S	9.84	16.00	12.25	6.60	0.95	M20
250 M	9.84	16.00	13.75	6.60	0.95	M20
280 S	11.02	18.00	14.50	7.50	0.95	M20
280 M	11.02	18.00	16.50	7.50	0.95	M20

Chart 45   IEC mounting dimensions—Foot-mounted AC machines dimensions for mounting.

IEC Mounting Dimensions

Foot-mounted AC Machines
Dimensions for Mounting* *(Continued)*

Frame Number	H	A	B	C	K	Bolt or Screw
315 S	12.40	20.00	16.00	8.50	1.11	M24
315 M	12.40	20.00	18.00	8.50	1.11	M24
355 S	13.97	24.00	19.70	10.00	1.11	M24
355 M	13.97	24.00	22.05	10.00	1.11	M24
355 L	13.97	24.00	24.80	10.00	1.11	M24
400 S	15.74	27.00	22.05	11.00	1.38	M30
400 M	15.74	27.00	24.80	11.00	1.38	M30
400 L	15.74	27.00	27.95	11.00	1.38	M30

IEC 72-1 Standards.
Dimensions, except for bolt and screw sizes, are shown in inches (rounded off). Bolt and screw sizes are shown in millimeters. For tolerances on dimensions, see IEC 72-1, 6.1, Foot-Mounted Machines, Table 1. (Note: Data in IEC tables is shown in millimeters.)

Chart 45 *Continued.*

IEC Shaft Extension, Key and Keyseat Dimensions

Continuous Duty AC Motors

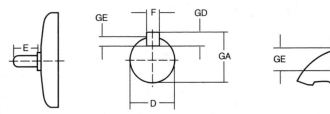

D	E	Key		Keyseat		GA	Max. Torque (Lb Ft)
		F	GD	F	GE		
0.2756	0.63	0.078	0.078	0.078	0.048	0.307	0.184
0.3543	0.79	0.118	0.118	0.118	0.071	0.401	0.465
0.4331	0.91	0.157	0.157	0.157	0.099	0.492	0.922
0.5512	1.18	0.196	0.196	0.196	0.119	0.629	2.07
0.6299	1.57	0.196	0.196	0.196	0.119	0.708	3.02
0.7087	1.57	0.236	0.236	0.236	0.138	0.807	5.24
0.7480	1.57	0.236	0.236	0.236	0.138	0.846	6.09
0.8661	1.97	0.236	0.236	0.236	0.138	0.964	10.3
0.9449	1.97	0.314	0.275	0.314	0.160	1.062	13.3
1.1024	2.36	0.314	0.275	0.314	0.160	1.220	23.2
1.2598	3.15	0.393	0.314	0.393	0.200	1.377	36.9
1.4961	3.15	0.393	0.314	0.393	0.200	1.614	66.4
1.6535	4.33	0.472	0.314	0.472	0.200	1.771	92.2
1.8898	4.33	0.551	0.354	0.551	0.220	2.027	148
2.1654	4.33	0.629	0.393	0.629	0.240	2.322	262
2.3622	5.51	0.708	0.433	0.708	0.280	2.519	332
2.5591	5.51	0.708	0.433	0.708	0.280	2.716	465
2.7559	5.51	0.787	0.472	0.787	0.300	2.933	590
2.9528	5.51	0.787	0.472	0.787	0.300	3.129	738
3.1496	6.69	0.866	0.551	0.866	0.355	3.346	922
3.3465	6.69	0.866	0.551	0.866	0.355	3.543	1180
3.5433	6.69	0.984	0.551	0.984	0.355	3.740	1401
3.7402	6.69	0.984	0.551	0.984	0.355	3.937	1696
3.9370	8.27	1.102	0.629	1.102	0.395	4.173	2065
4.3307	8.27	1.102	0.629	1.102	0.395	4.566	2950

IEC 72-1 Standards. All dimensions are rounded off. For tolerances on dimensions, see IEC 72-1, 7, Shaft Extension, Keys and Keyways Dimensions, Table 4. (Note: Data in IEC tables is shown in millimeters.)

Chart 46   IEC shaft extension, key and keyseat dimensions—Continuous duty AC motors.

IEC Mounting Dimensions*

Foot-Mounted AC Machines

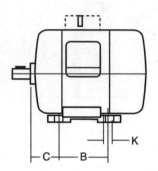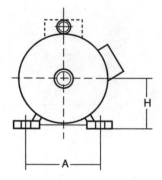

*Dimensions in millimeters

Frame Number	H	A	B	C	K	Bolt or Screw
56 M	56	90	71	36	5.8	M5
63 M	63	100	80	40	7	M6
71 M	71	112	90	45	7	M6
80 M	80	125	100	50	10	M8
90 S	90	140	100	56	10	M8
90 L	90	140	125	56	10	M8
100 S	100	160	112	63	12	M10
100 L	100	160	140	63	12	M10
112 S	112	190	114	70	12	M10
112 M	112	190	140	70	12	M10
132 S	132	216	140	89	12	M10
132 M	132	216	178	89	12	M10
160 S	160	254	178	108	14.5	M12
160 M	160	254	210	108	14.5	M12
160 L	160	254	254	108	14.5	M12
180 S	180	279	203	121	14.5	M12
180 M	180	279	241	121	14.5	M12
180 L	180	279	279	121	14.5	M12
200 S	200	318	228	133	18.5	M16
200 M	200	318	267	133	18.5	M16
200 L	200	318	305	133	18.5	M16

Chart 47 IEC mounting dimensions—Foot-mounted AC machines.

225 S	225	356	286	149	18.5	M16
225 M	225	356	311	149	18.5	M16
250 S	250	406	311	168	24	M20
250 M	250	406	349	168	24	M20
280 S	280	457	368	190	24	M20
280 M	280	457	419	190	24	M20
315 S	315	508	406	216	28	M24
315 M	315	508	457	216	28	M24
355 S	355	610	500	254	28	M24
355 M	355	610	560	254	28	M24
355 L	355	610	630	254	28	M24
400 S	400	686	560	280	35	M30
400 M	400	686	630	280	35	M30
400 L	400	686	710	280	35	M30

IEC 72-1 Standards. For tolerances on dimensions, see IEC 72-1, 6.1, Foot-Mounted Machines, Table 1.

Chart 47 *Continued.*

IEC Shaft Extension, Key and Keyseat Dimensions*

Continuous Duty AC Motors

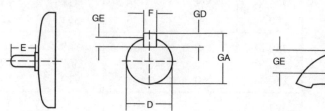

*Dimensions in millimeters

D	E	Key		Keyseat		GA	Max. Torque (NM)
		F	GD	F	GE		
7	16	2	2	2	1.2	7.8	0.25
9	20	3	3	3	1.8	10.2	0.63
11	23	4	4	4	2.5	12.5	1.25
14	30	5	5	5	3	16	2.8
16	40	5	5	5	3	18	4.1
18	40	6	6	6	3.5	20.5	7.1
19	40	6	6	6	3.5	21.5	8.25
22	50	6	6	6	3.5	24.5	14
24	50	8	7	8	4	27	18
28	60	8	7	8	4	31	31.5
32	80	10	8	10	5	35	50
38	80	10	8	10	5	41	90
42	110	12	8	12	5	45	125
48	110	14	9	14	5.5	51.5	200
55	110	16	10	16	6	59	355
60	140	18	11	18	7	64	450
65	140	18	11	18	7	69	630
70	140	20	12	20	7.5	74.5	800
75	140	20	12	20	7.5	79.5	1000
80	170	22	14	22	9	85	1250
85	170	22	14	22	9	90	1600
90	170	25	14	25	9	95	1900
95	170	25	14	25	9	100	2300
100	210	28	16	28	10	106	2800
110	210	28	16	28	10	116	4000

IEC 72-1 Standards. For tolerances on dimensions, see IEC 72-1, 7, Shaft Extension, Keys and Keyways Dimensions, Table 4.

Chart 48   IEC shaft extension, key and keyseat dimensions—Continuous duty AC motors.

FORMULAS

Miscellaneous Formulas

Ohms Law

Ohms = Volts/Amperes (R = E/I)
Amperes = Volts/Ohms (I = E/R)
Volts = Amperes × Ohms (E = IR)

Power—A-C Circuits

$$\text{Efficiency} = \frac{746 \times \text{Output Horsepower}}{\text{Input Watts}}$$

$$\text{Three-Phase Kilowatts} = \frac{\text{Volts} \times \text{Amperes} \times \text{Power Factor} \times 1.732}{1000}$$

$$\text{Three-Phase Volt-Amperes} = \text{Volts} \times \text{Amperes} \times 1.732$$

$$\text{Three-Phase Amperes} = \frac{746 \times \text{Horsepower}}{1.732 \times \text{Volts} \times \text{Efficiency} \times \text{Power Factor}}$$

$$\text{Three-Phase Efficiency} = \frac{746 \times \text{Horsepower}}{\text{Volts} \times \text{Amperes} \times \text{Power Factor} \times 1.732}$$

$$\text{Three-Phase Power Factor} = \frac{\text{Input Watts}}{\text{Volts} \times \text{Amperes} \times 1.732}$$

$$\text{Single-Phase Kilowatts} = \frac{\text{Volts} \times \text{Amperes} \times \text{Power Factor}}{1000}$$

$$\text{Single-Phase Amperes} = \frac{746 \times \text{Horsepower}}{\text{Volts} \times \text{Efficiency} \times \text{Power Factor}}$$

$$\text{Single-Phase Efficiency} = \frac{746 \times \text{Horsepower}}{\text{Volts} \times \text{Amperes} \times \text{Power Factor}}$$

$$\text{Single-Phase Power Factor} = \frac{\text{Input Watts}}{\text{Volts} \times \text{Amperes}}$$

$$\text{Horsepower (3 Phase)} = \frac{\text{Volts} \times \text{Amperes} \times 1.732 \times \text{Efficiency} \times \text{Power Factor}}{746}$$

$$\text{Horsepower (1 Phase)} = \frac{\text{Volts} \times \text{Amperes} \times \text{Efficiency} \times \text{Power Factor}}{746}$$

Power—D-C Circuits

Watts = Volts × Amperes (W = EI)

$$\text{Amperes} = \frac{\text{Watts}}{\text{Volts}} \ (I = W/E)$$

$$\text{Horsepower} = \frac{\text{Volts} \times \text{Amperes} \times \text{Efficiency}}{746}$$

Speed—A-C Machinery

$$\text{Synchronous RPM} = \frac{\text{Hertz} \times 120}{\text{Poles}}$$

$$\text{Percent Slip} = \frac{\text{Synchronous RPM} - \text{Full-Load RPM}}{\text{Synchronous RPM}} \cdot 100$$

Motor Application

$$\text{Torque (lb.-ft.)} = \frac{\text{Horsepower} \times 5250}{\text{RPM}}$$

$$\text{Horsepower} = \frac{\text{Torque (lb.-ft.)} \times \text{RPM}}{5250}$$

Time for Motor to Reach Operating Speed (Seconds)

$$\text{Seconds} = \frac{WK^2 \times \text{Speed Change}}{308 \times \text{Avg. Accelerating Torque}}$$

WK^2 = Inertia of Rotor + Inertia of Load (lb.-ft.2

$$\text{Average Accelerating torque} = \frac{[(FLT + BDT)/2] + BDT + LRT}{3}$$

FLT = Full-Load Torque   BDT = Breakdown Torque
LRT = Locked Rotor Torque

$$\text{Load } WK^2 \text{ (at motor shaft)} = \frac{WK^2 \text{ (Load)} \times \text{Load RPM}}{\text{Motor RPM}}$$

$$\text{Shaft Stress (P.S.I.)} = \frac{\text{HP} \times 321.000}{\text{RPM} \times \text{Shaft Diam.}}$$

Resistance = Temperature

$$R_C = R_H \times \frac{(K + T_C)}{(K + T_H)}$$

$$R_H = R_C \times \frac{(K + T_H)}{(K + T_C)}$$

K = 234.5 – Copper
   = 236 – Aluminum
   = 180 – Iron
   = 218 – Steel
R_C = Cold Resistance (°C)
R_H = Hot Resistance (°C)
T_C = Cold Temperature (°C)
T_H = Hot Temperature (°C)

Vibration

D = .318 (V/Hz)   D = Displacement (Inches Peak-Peak)
V = π (Hz) D   V = Velocity (Inches per Second Peak)
A = .051 (Hz) (D)   A = Acceleration (g's Peak)
A = .016 (Hz) (V)   Hz = Cycles per Second

Temperature Conversion Table

Locate known temperature in °C/°F column. Read converted temperature in °C or °F column.

°C	°C/°F	°F	°C	°C/°F	°F	°C	°C/°F	°F
−45.4	−50	−58	15.5	60	140	76.5	170	338
−42.7	−45	−49	18.3	65	149	79.3	175	347
−40	−40	−40	21.1	70	158	82.1	180	356
−37.2	−35	−31	23.9	75	167	85	185	365
−34.4	−30	−22	26.6	80	176	87.6	190	374
−32.2	−25	−13	29.4	85	185	90.4	195	383
−29.4	−20	−4	32.2	90	194	93.2	200	392
−26.6	−15	5	35	95	203	96	205	401
−23.8	−10	14	37.8	100	212	98.8	210	410
−20.5	−5	23	40.5	105	221	101.6	215	419
−17.8	0	32	43.4	110	230	104.4	220	428
−15	5	41	46.1	115	239	107.2	225	437
−12.2	10	50	48.9	120	248	110	230	446
−9.4	15	59	51.6	125	257	112.8	235	455
−6.7	20	68	54.4	130	266	115.6	240	464
−3.9	25	77	57.1	135	275	118.2	245	473
−1.1	30	86	60	140	284	120.9	250	482
1.7	35	95	62.7	145	293	123.7	255	491
4.4	40	104	65.5	150	302	126.5	260	500
7.2	45	113	68.3	155	311	129.3	265	509
10	50	122	71	160	320	132.2	270	518
12.8	55	131	73.8	165	329	136	275	527

°F = (9/5 × °C) + 32
°C = 5/9 (°F − 32)

QUICK REFERENCE TABLES

TABLE A   Insulation Resistance Tests on Electrical Apparatus and Systems

Maximum Voltage Rating of Equipment	Minimum Test Voltage, dc	Recommended Minimum Insulation Resistance in Megohms
250 Volts	500 Volts	25
600 Volts	1,000 Volts	100
5,000 Volts	2,500 Volts	1,000
8,000 Volts	2,500 Volts	2,000
15,000 Volts	2,500 Volts	5,000
25,000 Volts	5,000 Volts	20,000
35,000 Volts	15,000 Volts	100,000
46,000 Volts	15,000 Volts	100,000
69,000 Volts	15,000 Volts	100,000

NOTE: This table has recommended minimum insulation values identical to those in the NETA Acceptance Testing Specification for new equipment. Well maintained insulation in favorable ambient conditions should continue to provide these values.

See Table B for temperature correction factors.

TABLE B   **Insulation Resistance Conversion Factors
For Conversion of Test Temperature to 20° C**

Temperature		Multiplier	
°C	°F	Apparatus Containing Immersed Oil Insulations	Apparatus Containing Solid Insulations
0	32	0.25	0.40
5	41	0.36	0.45
10	50	0.50	0.50
15	59	0.75	0.75
20	68	1.00	1.00
25	77	1.40	1.30
30	86	1.98	1.60
35	95	2.80	2.05
40	104	3.95	2.50
45	113	5.60	3.25
50	122	7.85	4.00
55	131	11.20	5.20
60	140	15.85	6.40
65	149	22.40	8.70
70	158	31.75	10.00
75	167	44.70	13.00
80	176	63.50	16.00

TABLE C   **Maximum Allowable Vibration Amplitude**

Speed – RPM	Amplitude – Inches Peak to Peak
3000 and above	0.001
1500–2999	0.002
1000–1499	0.0025
999 and below	0.003

This table is from NEMA publication MG 1, Sections 20.53, 21.54, 22.54, 23.52, and 24.50.

TABLE D   Acceptable System Voltage Ranges

Nominal System Voltage	Allowable Limits %	Allowable Voltage Range
120 V (L – N)	± 5%	114 V – 126 V
240 V (L – L)	± 5%	228 V – 252 V
480 V (L – L)	± 5%	456 V – 504 V

Source: U.S. Department of Energy.

TABLE E   Characteristics of Motor Loads

Description of Motor Use	Type of Load
Centrifugal Supply Air Fan Motor	Constant, but will change slightly with outside air temperature.
Conveyor with Continuous and Constant Load	Constant
Boiler Feed Water Pump Motor, "On-Off" Control	Starts/stops. Constant while on.
Hydraulic Power Unit Motor, "On and Bypass" Control	Two levels of different but constant values.
Air Compressor Motor with "Low Unloaded hp" Control	Two Levels of different but constant values.
Air Compressor Motor with "Inlet Valve" Control	Random load.

Source: U.S. Department of Energy.

TABLE F   Induction Motor Synchronous Speeds

Poles	60 Hertz	50 Hertz
2	3,600	3,000
4	1,800	1,500
6	1,200	1,000
8	900	750
10	720	600
12	600	500

Source: U.S. Department of Energy.

TABLE G   Motor Loss Categories

No Load Losses	Typical Losses (%)	Factors Affecting these Losses
Core Losses	15–25	Type and quantity of magnetic material
Friction and Windage Losses	5–15	Selection and design of fans and bearings
Motor Operating Under Load		
Stator I^2R Losses	25–40	Stator conductor size
Rotor I^2R Losses	15–25	Rotor conductor size
Stray Load Losses	10–20	Manufacturing and design methods

Source: U.S. Department of Energy.

TABLE H   Typical Distributions of Motor Losses, % (1800 RPM Open Drip-Proof Enclosure)

Types of Loss	Motor Horsepower		
	25	50	100
Stator I^2R	42	38	28
Rotor I^2R	21	22	18
Core Losses	15	20	13
Windage and Friction	7	8	14
Stray Load	15	12	27

TABLE I   Efficiency Results from Various Motor Testing Standards

Standard	Full-Load Efficiency (%)	
	7.5 hp	20 hp
Canadian (CSA C390)	80.3	86.9
United States (IEEE-112, Test Method B)	80.3	86.9
International (IEC - 34.2)	82.3	89.4
British (BS - 269)	82.3	89.4
Japanese (JEC - 37)	85.0	90.4

Source: U.S. Department of Energy.

TABLE J Fan Laws/Affinity Laws

Law #1: $\dfrac{CFM_2}{CFM_1} = \dfrac{RPM_2}{RPM_1}$

Quantity (CFM) varies as fan speed (RPM)

Law #2: $\dfrac{P_2}{P_1} = \dfrac{(RPM_2)^2}{(RPM_1)^2}$

*Pressure (P) varies as the **square** of fan speed*

Law #3: $\dfrac{hp_2}{hp_1} = \dfrac{(RPM_2)^3}{(RPM_1)^3}$

*Horsepower (hp) varies as the **cube** of fan speed*

Source: U.S. Department of Energy.

TABLE K Sensitivity of Load to Motor RPM

$\dfrac{(1,760)^3}{(1,740)^3} = 3.5$ *percent horsepower increase*

$\dfrac{(1,780)^3}{(1,740)^3} = 7.0$ *percent horsepower increase*

TABLE L Grease Compatibility

	Aluminum Complex	Barium	Calcium	Calcium 12-hydroxy	Calcium Complex	Clay	Lithium	Lithium 12-hydroxy	Lithium Complex	Polyurea
Aluminum Complex	X	I	I	C	I	I	I	I	C	I
Barium	I	X	I	C	I	I	I	I	I	I
Calcium	I	I	X	C	I	C	C	B	C	I
Calcium 12-hydroxy	C	C	C	X	B	C	C	C	C	I
Calcium Complex	I	I	I	B	X	I	I	I	C	C
Clay	I	I	C	C	I	X	I	I	I	I
Lithium	I	I	C	C	I	I	X	C	C	I
Lithium 12-hydroxy	I	I	B	C	I	I	C	X	C	I
Lithium Complex	C	I	C	C	C	I	C	C	X	I
Polyurea	I	I	I	I	C	I	I	I	I	X

I = Incompatible   C = Compatible   B = Borderline
Source: U.S. Department of Energy.

Table M Table 430-150. Full-Load Current* Three-Phase
Alternating-Current Motors

HP	115V	Induction Type Squirrel-Cage and Wound-Rotor Amperes						Synchronous Type Unity Power Factor† Amperes			
		200V	208V	230V	460V	575V	2300V	230V	460V	575V	2300V
½	4	2.3	2.2	2	1	.8					
¾	5.6	3.2	3.1	2.8	1.4	1.1					
1	7.2	4.1	4.0	3.6	1.8	1.4					
1½	10.4	6.0	5.7	5.2	2.6	2.1					
2	13.6	7.8	7.5	6.8	3.4	2.7					
3		11.0	10.6	9.6	4.8	3.9					
5		17.5	16.7	15.2	7.6	6.1					
7½		25.3	24.2	22	11	9					
10		32.2	30.8	28	14	11					
15		48.3	46.2	42	21	17					
20		62.1	59.4	54	27	22					
25		78.2	74.8	68	34	27		53	26	21	
30		92	88	80	40	32		63	32	26	
40		119.6	114.4	104	52	41		83	41	33	
50		149.5	143.0	130	65	52		104	52	42	
60		177.1	169.4	154	77	62	16	123	61	49	12
75		220.8	211.2	192	96	77	20	155	78	62	15
100		285.2	272.8	248	124	99	26	202	101	81	20
125		358.8	343.2	312	156	125	31	253	126	101	25
150		414	396.0	360	180	144	37	302	151	121	30
200		552	528.0	480	240	192	49	400	201	161	40

*These values of full-load current are for motors running at speeds usual for belted motors and motors with normal torque characteristics. Motors built for especially low speeds or high torques may require more running current, and multispeed motors will have full-load current varying with speed, in which case the nameplate current rating shall be used.

†For 90 and 80 percent power factor, the above figures shall be multiplied by 1.1 and 1.25 respectively.

The voltages listed are rated motor voltages. The currents listed shall be permitted for system voltage ranges of 110 to 120, 220 to 240, 440 to 480, and 550 to 600 volts.

TABLE N Ampacity of Secondary Conductors for Wound Rotor Motors

Resistor Duty Classification	Ampacity of Conductor in Percent of Full-Load Secondary Current
Light starting duty	35
Heavy starting duty	45
Extra-heavy starting duty	55
Light intermittent duty	65
Medium intermittent duty	75
Heavy intermittent duty	85
Continuous duty	110

Reprinted with permission from NFPA 70—1996, the National Electrical Code®, Copyright © 1995, National Fire Protection Association, Quincy, MA 02269. This reprinted material is not the complete and official position of the National Fire Protection Association, on the referenced subject which is represented only by the standard in its entirety.

TABLE O Maximum Rating or Setting of Motor Branch-Circuit Short-Circuit and Ground-Fault Protective Devices

Type of Motor	Percent of Full-Load Current			
	Nontime Delay Fuse**	Dual Element (Time-Delay) Fuse**	Instantaneous Trip Breaker	Inverse Time Breaker*
Single-phase motors	300	175	800	250
AC polyphase motors other than wound-rotor Squirrel Cage:				
Other than Design E	300	175	800	250
Design E	300	175	1100	250
Synchronous†	300	175	800	250
Wound rotor	150	150	800	150
Direct-current (constant voltage)	150	150	250	150

For certain exceptions to the values specified, see Sections 430-52 through 430-54.

*The values given in the last column also cover the ratings of nonadjustable inverse time types of circuit breakers that may be modified as in Section 430-52.

**The values in the Nontime Delay Fuse Column apply to Time-Delay Class CC fuses.

†Synchronous motors of the low-torque, low-speed type (usually 450 rpm or lower), such as are used to drive reciprocating compressors, pumps, etc., that start unloaded, do not require a fuse rating or circuit-breaker setting in excess of 200 percent of full-load current.

Reprinted with permission from NFPA 70—1996, the National Electrical Code®, Copyright © 1995, National Fire Protection Association, Quincy, MA 02269. This reprinted material is not the complete and official position of the National Fire Protection Association on the referenced subject, which is represented only by the standard in its entirety.

TABLE P   Full-Load Current in Amperes, Direct-Current Motors

The following values of full-load currents* are for motors running at base speed.

HP	Armature Voltage Rating*					
	90V	120V	180V	240V	500V	550V
$\frac{1}{4}$	4.0	3.1	2.0	1.6		
$\frac{1}{3}$	5.2	4.1	2.6	2.0		
$\frac{1}{2}$	6.8	5.4	3.4	2.7		
$\frac{3}{4}$	9.6	7.6	4.8	3.8		
1	12.2	9.5	6.1	4.7		
$1\frac{1}{2}$		13.2	8.3	6.6		
2		17	10.8	8.5		
3		25	16	12.2		
5		40	27	20		
$7\frac{1}{2}$		58		29	13.6	12.2
10		76		38	18	16
15				55	27	24
20				72	34	31
25				89	43	38
30				106	51	46
40				140	67	61
50				173	83	75
60				206	99	90
75				255	123	111
100				341	164	148
125				425	205	185
150				506	246	222
200				675	330	294

*These are average direct-current quantities.

TABLE Q Full-Load Currents in Amperes Single-Phase Alternating-Current Motors

The following values of full-load currents are for motors running at usual speeds and motors with normal torque characteristics. Motors built for especially low speeds or high torques may have higher full-load currents, and multispeed motors will have full-load current varying with speed, in which case the nameplate current ratings shall be used.

The voltages listed are rated motor voltages. The currents listed shall be permitted for system voltage ranges of 110 to 120 and 220 to 240.

HP	115V	200V	208V	230V
$\frac{1}{6}$	4.4	2.5	2.4	2.2
$\frac{1}{4}$	5.8	3.3	3.2	2.9
$\frac{1}{3}$	7.2	4.1	4.0	3.6
$\frac{1}{2}$	9.8	5.6	5.4	4.9
$\frac{3}{4}$	13.8	7.9	7.6	6.9
1	16	9.2	8.8	8
$1\frac{1}{2}$	20	11.5	11	10
2	24	13.8	13.2	12
3	34	19.6	18.7	17
5	56	32.2	30.8	28
$7\frac{1}{2}$	80	46	44	40
10	100	57.5	55	50

TABLE R   Indication Type Squirrel Cage and Wound Rotor Amperes

HP	Induction Type Squirrel-Cage and Wound-Rotor Amperes				
	115V	230V	460V	575V	2300V
$\frac{1}{2}$	4	2	1	.8	
$\frac{3}{4}$	4.8	2.4	1.2	1.0	
1	6.4	3.2	1.6	1.3	
$1\frac{1}{2}$	9	4.5	2.3	1.8	
2	11.8	5.9	3	2.4	
3		8.3	4.2	3.3	
5		13.2	6.6	5.3	
$7\frac{1}{2}$		19	9	8	
10		24	12	10	
15		36	18	14	
20		47	23	19	
25		59	29	24	
30		69	35	28	
40		90	45	36	
50		113	56	45	
60		133	67	53	14
75		166	83	66	18
100		218	109	87	23
125		270	135	108	28
150		312	156	125	32
200		416	208	167	43

TABLE S1 **Conversion Table of Single-Phase Locked Rotor Currents for Selection of Disconnecting Means and Controllers as Determined from Horsepower and Voltages Rating**
For use only with Sections 430-110, 440-12, 440-41, and 455-8(c).

Rated HP	Maximum Locked-Rotor Current Amperes Single Phase		
	115 Volts	208 Volts	230 Volts
$\frac{1}{2}$	58.8	32.5	29.4
$\frac{3}{4}$	82.8	45.8	41.4
1	96	53	48
$1\frac{1}{2}$	120	66	60
2	144	80	72
3	204	113	102
5	336	186	168
$7\frac{1}{2}$	480	265	240
$10\frac{1}{2}$	600	332	300

TABLE S2   Conversion Table of Polyphase Design B, C, D, and E Maximum Locked-Rotor Currents for Selection of Disconnecting Means and Controllers as Determined from Horsepower and Voltage Rating and Design Letter

For use only with Sections 430-110, 440-12, 440-41, and 455-8(c).

Rated HP	Maximum Motor Locked-Rotor Current Amperes Two- and Three-Phase Design B, C, D, and E											
	115 Volts		200 Volts		208 Volts		230 Volts		460 Volts		575 Volts	
	B, C, D	E	B, C, D	E	B, C D	E	B, C, D	E	B, C, D	E	B, C, D	E
$\frac{1}{2}$	40	40	23	23	22.1	22.1	20	20	10	10	8	8
$\frac{3}{4}$	50	50	28.8	28.8	27.6	27.6	25	25	12.5	12.5	10	10
1	60	60	34.5	34.5	33	33	30	30	15	15	12	12
$1\frac{1}{2}$	80	80	46	46	44	44	40	40	20	20	16	16
2	100	100	57.5	57.5	55	55	50	50	25	25	20	20
3			73.6	84	71	81	64	73	32	36.5	25.6	29.2
5			105.8	140	102	135	92	122	46	61	36.8	48.8
$7\frac{1}{2}$			146	210	140	202	127	183	63.5	91.5	50.8	73.2
10			186.3	259	179	249	162	225	81	113	64.8	90
15			267	388	257	373	232	337	116	169	93	135
20			334	516	321	497	290	449	145	225	116	180
25			420	646	404	621	365	562	183	281	146	225
30			500	775	481	745	435	674	218	337	174	270
40			667	948	641	911	580	824	290	412	232	330
50			834	1185	802	1139	725	1030	363	515	290	412
60			1001	1421	962	1367	870	1236	435	618	348	494
75			1248	1777	1200	1708	1085	1545	543	773	434	618
100			1668	2154	1603	2071	1450	1873	725	937	580	749
125			2087	2692	2007	2589	1815	2341	908	1171	726	936
150			2496	3230	2400	3106	2170	2809	1085	1405	868	1124
200			3335	4307	3207	4141	2900	3745	1450	1873	1160	1498
250									1825	2344	1460	1875
300									2200	2809	1760	2247
350									2550	3277	2040	2622
400									2900	3745	2320	2996
450									3250	4214	2600	3371
500									3625	4682	2900	3746

APPENDIX

A

Motor Load and Operating Cost Analysis Form

Employee Name _____   Date _____

Company _____   Facility/Location _____

1. General Data

Application _____
What type of equipment the motor drives.

Energy rate _____ cents/kWh

Monthly Demand Rate _____

Annual Operating Hours _____

2. Motor Nameplate Data

Make/Model _____

Serial Number _____

Phase and Hz _____

Voltage Rating _____

Horsepower Rating _____

Enclosure Type _____

Synchronous Speed _____

Frame Size _____

NEMA Torque Class _____

Service Factor Rating _____

Insulation Class _____

Temperature Rise _____

FL Amperes Rating _____

FL RPM Rating _____

3. Measured Data

Input Volts _____
By Voltmeter

Input Amps _____
By Ampmeter

Input kW _____
By Powermeter if available, otherwise calculate below

Operating Speed _____
By Tachometer

4. Calculated Values

Full Load (FL) Slip _____
[Synchronous RPM – FL RPM Rating]

Operating Slip _____
[Synchronous RPM – Operating Speed]

Load Factor _____
[Operating Slip/FL Slip]

HP output _____
[Rated HP × Load Factor]

kVA input _____
[Input Volts × Input Amps × 0.001732]

Source: U.S. Department of Energy.

Power Factor _____
[(Average kW/kVA) × 100%, or based on Figure 2, using measured Input Amps/nameplate FL Amperage]

Input kW _____
[input Volts × input Amps × Power Factor × 0.001732]

Annual Operating Cost _____
[input kW × Annual Operating Hours × Energy Rate]

Annual Demand Cost (if any) _____
[Monthly Demand Rate × number of months motor operates during peak demand period]

Total Annual Operating Cost _____
[Annual Energy + Demand costs]

APPENDIX B

Efficiencies for 900 rpm, Standard Efficiency Motors						
Motor Size	ODP			TEFC		
Load Level	100%	75%	50%	100%	75%	50%
10	87.2%	87.6%	86.3%	86.8%	87.6%	86.8%
15	87.8%	88.8%	88.2%	87.5%	88.7%	88.1%
20	88.2%	89.2%	88.0%	89.2%	89.9%	89.2%
25	88.6%	89.2%	88.0%	89.7%	90.3%	89.1%
30	89.9%	90.7%	90.2%	89.6%	90.5%	86.5%
40	91.0%	91.8%	91.7%	90.5%	91.4%	85.5%
50	90.8%	91.9%	91.1%	90.2%	91.0%	90.2%
75	91.7%	92.4%	92.1%	91.6%	91.8%	91.0%
100	92.2%	92.2%	91.8%	92.4%	92.5%	92.0%
125	92.9%	92.3%	91.7%	93.0%	93.1%	92.1%
150	93.3%	93.1%	92.6%	93.0%	93.4%	92.5%
200	92.8%	93.5%	93.1%	93.7%	94.1%	93.4%
250	93.1%	93.5%	93.0%	91.7%	94.8%	94.5%
300	93.1%	93.7%	92.9%	94.4%	94.2%	93.7%

B1 Average efficiencies for standard and energy efficient motors at various load points; *Source:* U.S. Department of Energy.

Efficiencies for 1,200 rpm, Standard Efficiency Motors						
Motor Size	ODP			TEFC		
Load Level	100%	75%	50%	100%	75%	50%
10	87.3%	86.9%	85.7%	87.1%	87.7%	86.4%
15	87.4%	87.5%	86.8%	88.2%	88.1%	87.3%
20	88.5%	89.2%	88.8%	89.1%	89.7%	89.4%
25	89.4%	89.7%	89.3%	89.8%	90.5%	89.8%
30	89.2%	90.1%	89.8%	90.1%	91.3%	90.7%
40	90.1%	90.4%	90.0%	90.3%	90.1%	89.3%
50	90.7%	91.2%	90.9%	91.6%	92.0%	91.5%
75	92.0%	92.5%	92.3%	91.9%	91.6%	91.0%
100	92.3%	92.7%	92.2%	92.8%	92.7%	91.9%
125	92.6%	92.9%	92.8%	93.0%	93.0%	92.6%
150	93.1%	93.3%	92.9%	93.3%	93.8%	93.4%
200	94.1%	94.6%	93.5%	94.0%	94.3%	93.6%
250	93.5%	94.4%	94.0%	94.6%	94.5%	94.0%
300	93.8%	94.4%	94.3%	94.7%	94.8%	94.0%

B2 *Continued.*

Efficiencies for 1,800 rpm, Standard Efficiency Motors						
Motor Size	ODP			TEFC		
Load Level	100%	75%	50%	100%	75%	50%
10	86.3%	86.8%	85.9%	87.0%	88.4%	87.7%
15	88.0%	89.0%	88.5%	88.2%	89.3%	88.4%
20	88.6%	89.2%	88.9%	89.6%	90.8%	90.0%
25	89.5%	90.6%	90.0%	90.0%	90.9%	90.3%
30	89.7%	91.0%	90.9%	90.6%	91.6%	91.0%
40	90.1%	90.0%	89.0%	90.7%	90.5%	89.2%
50	90.4%	90.8%	90.3%	91.6%	91.8%	91.1%
75	91.7%	92.4%	92.0%	92.2%	92.5%	91.3%
100	92.2%	92.8%	92.3%	92.3%	92.1%	91.4%
125	92.8%	93.2%	92.7%	92.6%	92.3%	91.3%
150	93.3%	93.3%	93.0%	93.3%	93.1%	92.2%
200	93.4%	93.8%	93.3%	94.2%	94.0%	93.1%
250	93.9%	94.4%	94.0%	93.8%	94.2%	93.5%
300	94.0%	94.5%	94.2%	94.5%	94.4%	93.3%

B3 Comparison of purchase price versus running cost.
Source: U.S. Department of Energy.

Efficiencies for 3,600 rpm, Standard Efficiency Motors					
Motor Size	ODP			TEFC		
Load Level	100%	75%	50%	100%	75%	50%
10	86.3%	87.7%	86.4%	86.1%	87.2%	85.7%
15	87.9%	88.0%	87.3%	86.8%	87.8%	85.9%
20	89.1%	89.5%	88.7%	87.8%	89.6%	88.3%
25	89.0%	89.9%	89.1%	88.6%	89.6%	87.9%
30	89.2%	89.3%	88.3%	89.2%	90.0%	88.7%
40	90.0%	90.4%	89.9%	89.0%	88.4%	86.8%
50	90.1%	90.3%	88.7%	89.3%	89.2%	87.3%
75	90.7%	91.0%	90.1%	91.2%	90.5%	88.7%
100	91.9%	92.1%	91.5%	91.2%	90.4%	89.3%
125	91.6%	91.8%	91.1%	91.7%	90.8%	89.2%
150	92.0%	92.3%	92.0%	92.3%	91.7%	90.1%
200	93.0%	93.0%	92.1%	92.8%	92.2%	90.5%
250	92.7%	93.1%	92.4%	92.7%	92.5%	91.2%
300	93.9%	94.3%	93.8%	93.2%	92.8%	91.1%

B3 *Continued.*

APPENDIX

Purchase Price Versus Running Cost—A Comparison

Let's compare the fuel cost savings of an efficient automobile over a less efficient automobile with savings obtained from purchase of an energy-efficient over a standard-efficiency motor. Based upon 15,000 miles per year at a fuel economy of 25 miles per gallon with gasoline priced at $1.00 per gallon, the fuel cost of a typical car is $600 per year or about 6.6 percent of the $9,000 purchase price. A 5-mile-per-gallon improvement in fuel economy saves 100 gallons of gasoline valued at $100 annually.

In contrast, a 15-hp standard efficiency motor, continuously operating at 75 percent of its full rated load, would consume 85,189 kWh/year of electrical energy. At an electricity rate of only $.03/kWh, this energy is valued at $2,555 or 315 percent of the motor's list price. A typical 15-hp continuously operating energy-efficient motor conserves 3,863 kWh of electricity valued at $116 annually. Vehicle and motor purchase alternatives are summarized below.

Vehicle Versus Motor Purchase—Comparison Base Case

New Car (25 MPG)		New 15-hp Standard-Efficiency Motor	
Purchase Price:	$9,000	List Price:	$811
Drive:	15,000 miles/yr	Use:	Continuous
MPG:	25	Load Factor:	75 Percent
Gal/Yr:	600	kWh/Yr:	85,189
Fuel Cost Value @ $1.00/gal:	$600	Electricity Cost @ .03/kWh:	$2,555
Ratio of Annual Fuel Cost To Initial Cost:	6.6 Percent		315 Percent

Alternatives	
Fuel-Efficient Car (30 MPG)	**New 15-hp Energy-Efficient Motor**

Efficiency Improvement:	5 MPG	Efficiency Improvement: 86.3 to 90.4 percent	
Annual Fuel Savings:	100 Gal.	Annual kWh Savings:	3,863
Value of Savings:	$100/year	Value of Savings:	$116/year
Savings over 150,000 mile operating life: 1,000 gallons, $1,000		Savings over 20-year service life: 77,260 kWh and $2,320	

Over a 20-year operating period, the standard motor would consume approximately 1.7 million kWh of electrical energy. This energy is valued at $51,100 or more than 6,300 percent of the initial motor purchase price.

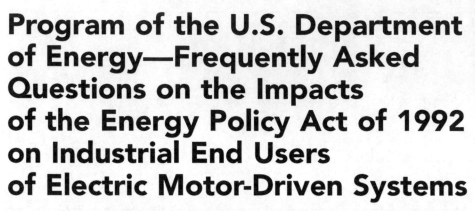

APPENDIX

D

Program of the U.S. Department of Energy—Frequently Asked Questions on the Impacts of the Energy Policy Act of 1992 on Industrial End Users of Electric Motor-Driven Systems

In general, what are the impacts of the Energy Policy Act of 1992 (EPACT) on end users of electric motor-driven systems?

Starting in October 1997, EPACT requires most general-purpose, polyphase, squirrel-cage, induction motors manufactured for sale in the United States rated 1 through 200 horsepower (hp) to meet minimum efficiency standards. Although many manufacturers now sell premium motors that meet these efficiency standards, most currently available motors do not. EPACT should increase the availability of energy-efficient motors for many end-use applications. In addition to motor efficiency standards, EPACT requires testing procedures and labeling.

To what kinds of electric motors will the new efficiency standards apply?

EPACT applies to general-purpose, T-frame, single-speed, foot-mounted, continuous-rated, polyphase, squirrel-cage, induction motors of National Electrical Manufacturers Association (NEMA) designs A and B. The subject motors are designed to operate on 230/460 volts and 60 hertz and have open and closed enclosures. EPACT applies to 6 pole (1200 rpm), 4 pole (1800 rpm), and 2 pole (3600 rpm) open and enclosed motors rated 1 through 200 hp. Such motors dominate industrial and commercial applications above 1 hp. EPACT does not apply to definite purpose motors (i.e., those designed for use under unusual conditions or for use on a particular type of application).

Will the new efficiency standards apply to existing motors?

EPACT applies to each motor manufactured (alone or as a component of another piece of equipment) after October 24, 1997. Existing motors and those manufactured between now and the implementation date are not governed by EPACT. Motors installed or in stock at industrial motor end-user facilities are unaffected.

What are the electric motor efficiency levels prescribed in the EPACT and how were they determined?

The table on page 292, which appears in EPACT, shows the required energy efficiency levels. EPACT prescribes nominal full-load efficiency levels based on horsepower ratings and enclosure types consistent with the suggested "Efficiency Levels of Energy-Efficient Polyphase Squirrel-Cage Induction Motors," in NEMA Standards Publication MG1-1987, "Motors and Generators," paragraph MG1-12.55A, Table 12-6C, which were in effect in 1992.

Are any motors currently being manufactured in compliance with the new efficiency standards?

Most motor manufacturers' current top-of-the-line energy-efficient motors meet the new efficiency levels. For some motor sizes and enclosures, a majority of manufacturers' current line of high efficiency motors meets the standards, although only a few manufacturers' motors currently meet the standards for other sizes. Almost without exception, currently available standard efficiency motors do not meet the new efficiency standards.

Who will test and certify the efficiency of motors?

Currently, manufacturers test and certify their own motors using the Institute of Electrical and Electronics Engineers and NEMA standards. NEMA is working in conjunction with the National Voluntary Laboratory Accreditation Program to establish an accreditation program for laboratories pursuant to a proposed DOE test procedure. Under a rule being considered by DOE, a manufacturer could certify that its motors meet efficiency levels prescribed by EPACT through testing conducted either by an accredited independent laboratory, or by the manufacturer's own laboratory (if it is accredited), or through a third-party certification organization that is nationally recognized. EPACT does not specify a timetable for DOE to require manufacturer laboratory certification or testing by independent laboratories.

Do the efficiency standards apply to imported motors and motors purchased as components of industrial equipment?

The standards apply to both imported motors and motors purchased as components of equipment. For example, general-purpose, polyphase, induction motors in packaged air compressor systems would be covered. Motors made in the United States for export are not subject to the standards.

Does EPACT prescribe overall efficiency levels for entire electric motor-driven systems used in manufacturing, such as packaged compressed air systems?

EPACT does not prescribe overall efficiency standards for any traditional industrial electric motor-driven systems, although it does contain efficiency standards for electric motor-driven systems such as air conditioners, which are used in the residential, commercial, and industrial sectors.

Electric Motor Efficiency Levels Prescribed in the Energy Policy Act of 1992
Nominal Full-Load Efficiency, Open Motors and Enclosed Motors.

Number of Poles	6	4	2	6	4	2
Motor Horsepower						
1	80.0	82.5	—	80.0	82.5	75.5
1.5	84.0	84.0	82.5	85.5	84.0	82.5
2	85.5	84.0	84.0	86.5	84.0	84.0
3	86.5	86.5	84.0	87.5	87.5	85.5
5	87.5	87.5	85.5	87.5	87.5	87.5
7.5	88.5	88.5	87.5	89.5	89.5	88.5
10	90.2	89.5	88.5	89.5	89.5	89.5
15	90.2	91.0	89.5	90.2	91.0	90.2
20	91.0	91.0	90.2	90.2	91.0	90.2
25	91.7	91.7	91.0	91.7	92.4	91.0
30	92.4	92.4	91.0	91.7	92.4	91.0
40	93.0	93.0	91.7	93.0	93.0	91.7
50	93.0	93.0	92.4	93.0	93.0	92.4
60	93.6	93.6	93.0	93.6	93.6	93.0
75	93.6	94.1	93.0	93.6	94.1	93.0
100	94.1	94.1	93.0	94.1	94.5	93.6
125	94.1	94.5	93.6	94.1	94.5	94.5
150	94.5	95.0	93.6	95.0	95.0	94.5
200	94.5	95.0	94.5	95.0	95.0	95.0

When will the new efficiency standards go into effect?

Covered motors manufactured (alone or as a component of another piece of equipment) in or imported into the United States after October 24, 1997, must meet the efficiency standards prescribed by EPACT. All motors manufactured or imported before that date can still be legally sold. Motors that require listing or certification with safety testing laboratories have an additional 2 years to meet the efficiency standards. In addition, small manufacturers of motors (less than $8 million in gross revenue) may petition for a 2-year exemption. After 2 years, the efficiency levels will be reevaluated by DOE to determine if more stringent standards should be developed and implemented.

Does the Energy Policy Act of 1992 give DOE the authority to prescribe any additional efficiency standards for electric motors?

EPACT provides DOE with the authority to prescribe efficiency standards and testing procedures for small horsepower motors if DOE determines the standards would be technologically feasible and economically justifiable and would result in significant energy savings. DOE must first prescribe test procedures for small motors, and then, within 18 months of the test procedures, prescribe efficiency standards. At this time, no test procedures or standards have been prescribed for small electric motors. EPACT also gives DOE the authority to fund programs for advanced electric motor system technologies.

Will local and state regulations be affected by the new efficiency standards?

After the implementation of the efficiency standards prescribed by EPACT, state and local regulations concerning energy efficiency or energy use by motors will be preempted.

How else will the Energy Policy Act of 1992 affect industrial end users of motors?

EPACT should increase the range of energy-efficient motors that are available in the marketplace. Users should consult with manufacturers to determine if new energy-efficient motors meet their horsepower, speed, and torque requirements. At the same time, DOE is planning a number of initiatives to promote the use of energy-efficient industrial motor-driven systems through the Motor Challenge, which is a voluntary collaboration between DOE and stakeholders in electric motor-driven systems.

How can I get more information on the impacts of the Energy Policy Act of 1992 on electric motor end users and the Motor Challenge?

By becoming a Partner in the Department of Energy's Motor Challenge, industrial motor end users can keep up to date on the latest developments concerning the Energy Policy Act of 1992 and receive other reliable information to enhance the quality and profitability of electric motor-driven system strategies and decisions.

About Motor Challenge

Motor Challenge is a partnership program between the U.S. Department of Energy and the nation's industries. The program is committed to increasing the use of energy-efficient, industrial electric motor systems and related technologies.

The program is wholly funded by the U.S. Department of Energy and is dedicated to helping industry increase its competitive edge, while conserving the nation's energy resources and enhancing environmental quality.

For More Information

Contact the Motor Challenge Information Clearinghouse: 1-800-862-2086. The Motor Challenge Information Clearinghouse is your one-stop resource for objective, reliable, and timely information on electric motor-driven systems.

Example 1A—Descriiption and Code (NEC) References for Motor Electrical Calculations

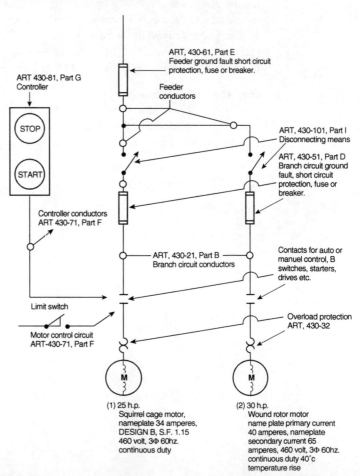

ART, 430-61, Part E
Feeder ground fault short circuit protection, fuse or breaker.

Feeder conductors

ART 430-81, Part G
Controller

STOP

START

ART, 430-101, Part I
Disconnecting means

ART, 430-51, Part D
Branch circuit ground fault, short circuit protection, fuse or breaker.

Controller conductors
ART 430-71, Part F

ART, 430-21, Part B
Branch circuit conductors

Contacts for auto or manuel control, B switches, starters, drives etc.

Limit switch

Overload protection
ART, 430-32

Motor control circuit
ART-430-71, Part F

M

M

(1) 25 h.p.
Squirrel cage motor, nameplate 34 amperes, DESIGN B, S.F. 1.15 460 volt, 3Φ 60hz. continuous duty

(2) 30 h.p.
Wound rotor motor name plate primary current 40 amperes, nameplate secondary current 65 amperes, 460 volt, 3Φ 60hz. continuous duty 40°c temperature rise

Example 1B—Calculation

Problem: Size motor conductors and overcurrent protection for example 1A.

Step 1 Size branch circuit conductors

- FLA-based on art 430–22, Table 430–150, and 430–6(a)
  A) FLA for the 25 h.p. motor = 34 amps.
  B) FLA for the 30 h.p. motor = 40 amps.
  C) Multiply FLA by 125% based on Art 430–22
    1) 25 h.p.—34A × 1.25 = 42.5 amperes
    2) 30 h.p.—40× 1.25 = 50 amperes
- Size branch circuit conductors–Art, 310–16, based on 75° C column
  A) 25 h.p.—42.5 FLA = #8 cu., 75° C, Table 310–16
  B) 30 h.p.—50 FLA = #8 cu., 75° C, Table 310–16

Step 2 Size feeder circuit conductors

Art 430–24, 310–16

- Ampacity of the conductors is based on Art 430–24. Conductor ampacity shall be the FLA sum of all motors plus (+) 25% of the highest rated motor in the group.
  A) The feeder ampacity is:
    1) 40 FLA (30 hp) × 125% + 40 FLA (30 hp) +
    2) 34 FLA (25 hp) = 124 amperes
  B) Conductor size is:
    1) #1 cu. based on 75° C, Table 310–16

Step 3 Size overload protection

Art 430–32(a)(1), 430–6(a)

- Size overload protection based on name plate FLA., temperature rise (40° C), service factor (1.15), FLA not greater than 9 amps, FLA 9.1 amps to 20 amps, FLA greater than 20 amperes, 1 hp or less (non-auto start), 1 hp or less (automatic start)
  A) Overload protection is:
    1) 25 hp–34 FLA (Nameplate) × 125% (Service factor 1.15, Table 430–32(1)(a)) = 42.5 amps
    2) 30 hp–40 FLA (Nameplate) × 125% (40° C temp. rise) = 50 amps
  B) If the motor overload protection is not sufficient to start the motor or carry the load, it is permitted to be increased in accordance with art. 430–34.
  C) Thermal protection (internal overload protection) can be used in lieu of separate overloads in accordance with art 430–7(a) 13, 430–32(a)(2).

Step 4 Branch circuit, short circuit and ground fault protection
Art 430–51, part D, 240–3, 6

- Overcurrent protection is based on table 430–152.

The following are exceptions, art 430–52(c)(1):

A) When calculated size overcurrent protection does not correspond to standard size the next higher standard size or setting can be used.

B) When calculated size will not allow motor to start the following can be applied.

1) Not greater than 600 volts or time delay class cc fuse can be increased by 400% of the FLA.

2) The rating of a dual element time delayed fuse can be increased by not greater than 225% of FLA.

3) Inverse time circuit breakers can be increased by 400% for protection 100 amps or less and 300% for protection greater than 100 amps.

4) FLA protection can be increased by 300% for fuses 601–6000 amperes.

5) Manufacturers specifications for protection cannot be exceeded.

6) Instantaneous trip breakers can be used only if adjustable and part of a combination listed controller. In this case a multiplier of 1300% of FLA shall be allowed to start motor.

7) When using instantaneous trip circuit breakers an increase of (A) 1300% of FLA for other than design E motors, (B) no more than 1700% for design E motors, (C) 800% for other than design E motors under engineering evaluation, (D) 1100% for design E motors under engineering evaluation, can be used when calculations based on Table 430-152 do not allow the motor to start.

8) Listed combination motor controllers rated at less than 8 amperes with an instantaneous-trip circuit breaker and a continuous rating of 15 amperes can be increased by the value marked on the controller.

9) One overcurrent protection for the smallest winding can be used in multi-speed applications.

10) Torque motors can be protected with the name plate FLA instead of Table 430–150.

C) 25 hp—34 amps (Table 430–150) × 300% (Table 430–152 nontime-delay fuse) = 102 amperes

1) The next higher size fuse is 110 amperes (240–6, 430–52(c) exception no. 1)

D) 30 hp—40 amps (Table 430–150) × 175% (Table 430–152 time delay fuse) = 70 amps fuse 430–52(c) exception no. 2A

1) 70 ampere fuse. (art 240–6)

Example 2

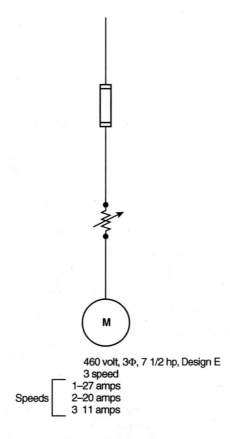

460 volt, 3Φ, 7 1/2 hp, Design E
3 speed
1–27 amps
Speeds   2–20 amps
3 11 amps

Step 5 Size feeder overcurrent protection

- To size the feeder O.C.P.D. we take the largest OCPD (fuse or breaker) plus(+) the sum of the full load currents of all other motors.

  A) Size feeder overcurrent protection is:

      1) 100A + 40 + 40 = 190 amperes.

      2) The nearest standard fuse is 175 amperes art 240–6, 430–62(a)

- Based conductor size and overcurrent protection on the largest full load amperage.

  A) Conductor size, branch circuit—art. 430–22(a)

      1) 125% × 27a = 33.75 ampere

  Table 310–16 75° C Conductor = #10 cu.

  B) Overcurrent protection, branch circuit—art. 430–52(a) excp

      1) Time delay fuse is:

         27 FLA (Table 430-150) × 175% (Table 430–152) = 47.25 amperes 50A time delay fuse

Example 3

25 h.p., 460 volt, 3Φ, cont. duty, Design B

- The disconnect must be horsepower rated switch, circuit breaker or a molded case switch.

  A) Size disconnecting means, art. 430–110 115% (430–110) × 34A (Table 430–150) = 39.1 amps. for non motor loads

  B) For combination loads use Table T or Tables 151a & b of the N.E.C.

    1) For general use a snap switches can be used as a disconnecting means for motors rated 2 horsepower or less.

    2) Cord and plug connected motors 1/3 horsepower or less can be used as a disconnecting means. See Figures 2–6, 2–7, & 2–8 for disconnect types.

Example 4

Effect of Voltage Unbalance on Motor Performance

When the line voltages applied to a polyphase induction motor are not equal, unbalanced currents in the stator windings will result. A small percentage voltage unbalance will result in a much larger percentage current unbalance. Consequently, the temperature rise of the motor operating at a particular load and percentage voltage unbalance will be greater than for the motor operating under the same conditions with balanced voltages.

Should voltages be unbalanced, the rated horsepower of the motor should be multiplied by the factor shown in the graph below to reduce the possibility of damage to the motor. Operation of the motor at above a 5 percent voltage unbalance condition is not recommended.

Alternating current, polyphase motors normally are designed to operate successfully under running conditions at rated load when the voltage unbalance at the motor terminals does not exceed 1 percent. Performance will not necessarily be the same as when the motor is operating with a balanced voltage at the motor terminals.

Medium Motor Derating Factor Due to Unbalanced Voltage

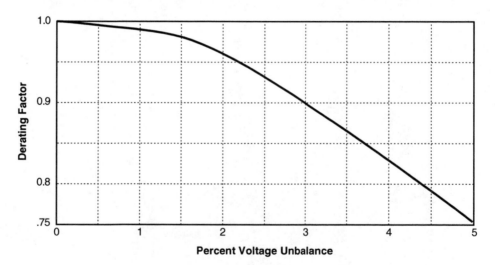

$$\text{Percent Voltage Unbalance} = 100 \times \frac{\text{Max. Volt. Deviation from Avg. Volt.}}{\text{Average Volt.}}$$

Example: With voltages of 460, 467, and 450, the average is 459, the maximum deviation from the average is 9, and the

$$\text{Percent Unbalance} = 100 \times \frac{9}{459} = 1.96 \text{ percent}$$

Reference: NEMA Standards MG 1-14.35.

INDEX